166 Topics in Current Chemistry

W0107413

Computer Chemistry

Editor: I. Ugi

With contributions by
K. Bley, J. Brunvoll, R. Carlson, B. N. Cyvin,
S. J. Cyvin, B. Gruber, E. Hladka, M. Knauer,
J. Koca, M. Kratochvil, V. Kvasnicka, L. Matyska,
A. Nordahl, J. Pospichal, V. Potucek, N. Stein, I. Ugi

With 67 Figures and 34 Tables

Springer-Verlag Berlin Heidelberg GmbH

This series presents critical reviews of the present position and future trends in modern chemical research. It is addressed to all research and industrial chemists who wish to keep abreast of advances in their subject.

As a rule, contributions are specially commissioned. The editors and publishers will, however, always pleased to be receive suggestions and supplementary information. Papers are accepted for "Topics in Current Chemistry" in English.

ISBN 978-3-662-14929-4   ISBN 978-3-540-47308-4 (eBook)
DOI 10.1007/978-3-540-47308-4

Library of Congress Catalog Card Number 74-644622

© Springer-Verlag Berlin Heidelberg 1993
Originally published by Springer-Verlag Berlin Heidelberg New York in 1993
Softcover reprint of the hardcover 1st edition 1993
The use of general descriptive names, registered namen, trademarks, etc. in this publication does not imply, even in the absence of a specific statement, that such names are exempt from the relevant protective laws and regulations and therefore for general use.

Typesetting: Th. Müntzer, Bad Langensalza;

51/3020-5 4 3 2 1 0 — Printed on acid-free paper

Guest Editor

Prof. Dr. *Ivar Ugi*
Organisch-Chemisches Institut
Technische Universität München
Lichtenbergerstraße 4, W-8046 Garching

Editorial Board

Preface

The considerable progress in chemistry made recently is due to the systematic analysis, classification, generation and processing of chemical information. The importance of abstract models, mathematical theories and computer-assistance for chemistry is increasingly recognised.

Mathematics has entered chemistry via physics. Many quantitative mathematical theories of chemistry, like quantum chemistry, are physical or physico-chemical in nature. Such theories play an important role in the design and evaluation of experiments, and they are used to compute numerical values of some observable properties of well-defined molecular systems. The results that are obtained through physics-based theories are indispensable to the indirect solution of a great variety of chemical problems.

In chemometrics, quantitative mathematical methods are used in the direct solution of chemical problems without the intermediacy of physics. There are also direct qualitative mathematical approaches to chemistry. More than a hundred years ago, Cayley demonstrated the usefulness of graph theory for representing the chemical constitution of molecules. The past decades have seen a vigorous renaissance of chemical applications of graph theory. Balaban's and Harary's initiative played a particularly important role in this resurgence. Pólya's group theoretical enumeration of isomers was the beginning of the wide field of direct uses of group theory in chemistry. The direct applications of graph theory and group theory are now the essential contents of the rapidly expanding discipline of mathematical chemistry. This discipline mainly deals with the static aspect of chemistry; it concentrates on the description, classification and enumeration of individual static chemical objects.

About 20 years ago, the theory of the be- and r-matrices, a global algebraic model of the logical structure of constitutional chemistry was formulated. This theory is the first direct mathematical approach to chemistry which also accentuates its dynamic aspect. The representation by mathematics comprises the individual objects of chemistry and also their relations, including their interconvertibility by chemical reactions. A decade later, the theory of the chemical identity groups was published in a monograph. It is a unified theory of stereochemistry that is primarily devoted to relations between molecular systems.

These two publications indicated the importance and feasibility of mathematical approaches that describe the relations between molecular systems. They stimulated the formulation of further mathematical models and theories of chemistry that are of interest in their own right but are particularly useful for computer-assistance in chemistry.

In general, the representation of chemical objects, facts and processes by mathematics not only improves the interpretation of chemistry but often it also forces chemists to express statements and definitions with greater precision and to specify exactly the conditions and range of their validity. It is increasingly recognised, that it is not only inevitable to use mathematical formalisms as quantitative theories and as a basis of numerical computations, but also as a foundation of computer programs for the qualitative solution of chemical problems such as the design of syntheses. The answers to the latter type of problems are molecules and chemical reactions.

When this kind of computer-assistance is mathematically-based and does not rely on detailed empirical chemical information, it can reach beyond the horizon of known chemistry, and the desirable solutions of given problems can be picked from the wealth of conceivable results by formalized selection procedures without the lottery of heuristic selection procedures. Appreciable advances in chemistry can be expected from further combinations of new concepts and models with mathematical formalisms and formal algorithms.

This issue of Topics in Current Chemistry contains four articles whose common denominator is the computer-oriented use of abstract concepts and mathematics in chemistry. The present articles will, hopefully, not only spread information about these types of endeavours but also stimulate some interest therein.

The contribution of Cyvin, Cyvin and Brunvoll on the enumeration of benzenoid chemical isomers is the continuation of a series that began with two articles in Volume 162 of this journal. These three articles constitute a definitive and exhaustive treatment of the given topic.

Carlson and Nordahl present a comprehensive account on the screening and optimization of organic syntheses by computer-assisted multivariate statistical methods. This article is rather an introduction to a philosophy than the description of a computer-assisted methodology. Thus it is particularly useful to those who are trying to understand the principles of this increasingly important technique to optimise syntheses.

In the past decade Kvasnička and his colleagues at Bratislava and Brno have published a series of graph-theory-based contributions to mathematical chemistry. The underlying concept is related to the algebra of be- and r-matrices. The present paper describes the Synthon Model of organic chemistry and the corresponding synthesis design program PEGAS. It is part of the aforementioned series.

The fourth paper in this volume is devoted to some extensions and generalizations of the algebra of the be- and r-matrices. The latter is only valid for the chemistry of molecular systems that are representable by integer bond orders and thus is not applicable to the great variety of molecules with multi-center bonds and delocalized electron systems. This deficiency is overcome by the introduction of the so-called extended be- and r-matrices the xbe- and xr-matrices. They contain additional rows/columns which refer to the delocalized electron systems. Some corresponding data structures are presented that also account for stereochemical aspects.

München, October 1992   Ivar Ugi

Attention all "Topics in Current Chemistry" readers:

A file with the complete volume indexes Vols. 22 (1972) through 163 (1992) in delimited ASCII format is available for downloading at no charge from the Springer EARN mailbox. Delimited ASCII format can be imported into most databanks.

The fie has been compressed using the popular shareware program "PKZIP" (Trademark of PKware Inc., PKZIP is available from most BBS and shareware distributors).

This file is distributed without any expressed or implied warranty.

To receive this file send an e-mail message to:
SVSERV@DHDSPRI6.BITNET.
The message must be: "GET /TCC/TCC‑CONT.ZIP".

SPSERV is an automatic data distribution system. It responds to your message. The following commands are available:

HELP	returns a detailed instruction set for the use of SVSERV,
DIR *(name)*	returns a list of files available in the directory "name",
INDEX *(name)*	same as "DIR",
CD ‹*name*›	changes to directory "name",
SEND‹*filename*›	invokes a message with the file "filename",
GET ‹*filename*›	same as "SEND".

Table of Contents

Exploring Organic Synthetic
Experimental Procedures

Rolf Carlson and Åke Nordahl

Department of Organic Chemistry, Umeå University, S-90187 Umeå, Sweden

Table of Contents

Topics in Current Chemistry, Vol. 166
© Springer-Verlag Berlin Heidelberg 1993

1 Introduction

Computers are playing increasingly important roles in chemical research. They are used for many different purposes such as measuring and control, signal processing, molecular modelling, quantum mechanical calculations, etc. In this chapter, we will discuss methods for design and evaluation of *experiments* in organic synthetic chemistry. All experimentally determined data are afflicted by an error component and as this also applies to data recorded in synthetic experiments, it is necessary to use statistical principles for distinguishing systematic variations induced by varying the experimental conditions from random error noise. Chemical phenomena rarely depend on single experimental factors and as experimental factors often also show interaction effects it will be necessary to run experiments which take all pertinent factors into account simultaneously. This calls for multivariate statistical methods. Statistics sometimes involve extensive computations and applications to organic synthesis are no exceptions. For this, computers are invaluable tools. The methods discussed in this chapter are not too demanding and can easily be handled on personal computers using commercially available software. The strategies outlined below should be seen as a methodology for problem solving in organic synthesis. As such it presents primarily a philosophical approach and not explicitly a computer methodology. Some of the methods described in this chapter have actually been developed B.C. (before computers) and involve rather trivial calculations and, hence, do not require computers for simple applications. The use of computers will, however, allow a more thorough and efficient examination of the results including different types of graphical representations which facilitate evaluations.

The methods described below are based on well-established statistical principles and some methods have been applied to problems in many different branches of science for many years while other methods are of a more recent date. To our experience, knowledge of multivariate statistical methods is, however, not widely spread in the community of organic chemical research and this chapter can be regarded as an introduction to the use of such methods in organic synthetic experimentation. We attempt to give an overview which explains the underlying principles and highlights the methods by selected examples. It is not possible to give an extensive review which covers the entire topic in one single chapter. A text-book which gives a thorough treatment of the subject has been published [1].

The examples shown have largely been obtained in our own work. This is not to promote our own results, there is another reason. In the examples from our laboratory, the reasoning behind the experiments is known in detail, whereas in published examples from other sources, certain details are sometimes not clearly stated and it is difficult to trace the logic behind the experiment.

2 Problem Area

The objectives of organic synthetic chemistry are to establish efficient methods for the synthesis of desired target molecules and in this respect, organic synthesis is an applied science. Often, the target molecules must be prepared by multi-step synthesis in which the final structure is built from smaller building blocks. The armoury of synthetic chemistry contains the combination of all known chemical compounds and all known chemical reactions. The number of such combinations is overwhelmingly large and it is impossible for any human being to grasp more than a tiny fraction of these. Computers may help in this situation, and there are several computerized retrieval systems avaliable, e.g. Chemical Abstracts (CAS on line), Beilstein's Handbook of Organic Chemistry, Theilheimer's Synthetic Methods, REACCS. This is, however, a rather unsophisticated application of computers. More elaborate applications are found in the strategic programs by which viable synthetic pathways from avalable starting materials to the desired target can be searched. Extensive reviews have been presented [2], see also the chapter by Kvasnicka in this volume. The majority of these programs for finding synthetic paths utilizes libraries of known chemical reactions. Exceptions are the approaches taken by Ugi and by Zefirov which use topological descriptions of molecules to define the bonds being made or broken. These programs are open-ended and do not rely on databases of known reactions [3, 4].

All strategies for multistep synthesis depend on the results obtained in the individual synthetic reaction steps. A necessary requisite for a successful total synthesis is therefore to have access to efficient procedures for carrying out all the intermediary steps. To ensure an optimum result, each reaction involved must be adjusted towards an optimum performance.

2.1 Optimization

The word "optimization" is used in many contexts in chemistry. In organic synthesis it usually means the adjustment of experimental conditions (pH, temperature, concentrations of reagents, etc) to achieve, e.g. a maximum yield. Sometimes the term denotes the adjustment of the reaction system (type of reagent, nature of solvent, etc) to achieve an efficient transformation. In this chapter, we will use the term to denote "a systematic search for improvement".

When a chemical *reaction* is elaborated into a synthetic *method* a number of questions arise which must be answered before an optimum method can be established:

— Which experimental variables have a significant influence and for this reason must be carefully controlled?
— How should the important experimental variables be adjusted to ensure an optimum result?
— Is it possible to apply the method to a variety of similar substrates? What are the scope and limitations of the reaction in this respect?

- Will it be possible to use different types of solvents for carrying out the transformation so that the reaction can be combined with other transformations in one-pot procedures? What is the scope of the reaction with regard to solvent variation?
- Is it possible to use different reagents to achieve the desired transformation? How should the reagent be modified to ensure an optimum result?
- Which properties of the reaction system (nature of substrate, type of reagent(s), type of solvent) are important for achieving a selective reaction?
- How will the optimum experimental conditions change if the structure of the substrate is changed?

When plausible reaction mechanisms are known, it is sometimes possible to answer some of the above questions by theoretical considerations. With new chemical reactions, however, a detailed reaction mechanism is not known and it is not likely that such details will be revealed before the utility of the reaction has been demonstrated. This means the optimization must often be preceded by a thorough mechanistic understanding which implies that the questions above must be answered through inferences from experimental observations. This imposes requirements on the experimental design, i.e. how different perturbations of the reaction conditions should be introduced and how the results thus obtained should be examined to furnish the desired information. In this chapter, we shall discuss various aspects of this topic.

2.2 Theme and Variations

A very popular form of music at the end of the 18th and the beginning of the 19th century was the "Theme and variation", and we use this as a metaphor to describe the problems encountered when a synthetic reaction is studied. The "theme", which applies to any synthetic reaction, would be

$$\text{Substrate} \xrightarrow[\text{Solvent}]{\text{Reagent}} \text{Product}$$

This "theme" can be varied in a number of ways:

The reaction may be applied to a variety of substrates and may have a broad scope in this respect. Reagents can be modified in a number of ways. An extreme example of this is furnished by the complex hydrides. A change in solvent may alter the reactivity pattern which may improve the result. Variations like these constitute discrete changes of the reaction system. The union of all such possible discrete variations of a given reaction is called the *reaction space*, see Fig. 1. The system variables are discrete and lack a single measure which describes their state. For the moment it is sufficient to regard the "axes" of the reaction space as mere variations. We shall see below that it is possible to quantify these axes through the concept of *principal properties*. This offers a means for systematically exploring the reaction space.

The situation is further complicated since in addition to the possible discrete changes, each reaction system can be subjected to innumerable variations of the detailed experimental conditions. These conditions are specified by the settings of continuous experimental variables like temperature, concentration of the various species in the reaction mixture, flow rates, etc. The experimental variables in this context are those factors which are controlled by the experimenter and which can be set to predetermined levels in the actual experiment. The union of all possible variations of the experimental variables is called the *experimental space*.

The observed result of an experimental called the *response*. The yield of the desired product is often the most important response and is sometimes used as a single criterion of success. It is, however, common that several responses are of interest, e.g. yields of byproduct, selectivity, cost for producing a given quantify of product. When several responses are considered we will also have to consider variations in a multidimensional *response space*, see Fig. 1.

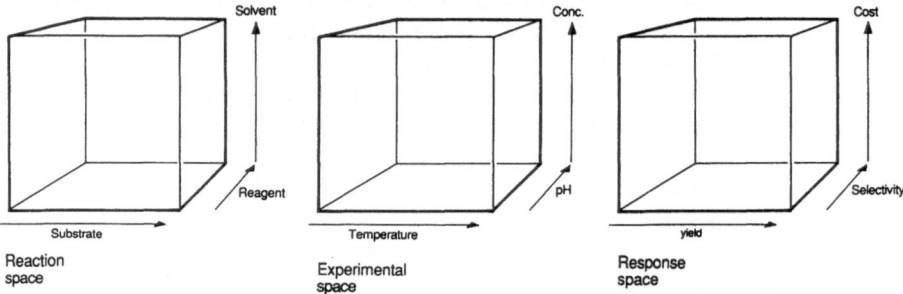

Fig. 1. Variations of synthetic reactions are illustrated by the reaction space, the experimental space, and the response space

This means that optimization in the general case will imply the exploration of the reaction space with a view to finding a suitable reaction system for carrying out the desired transformation followed by adjustment of the experimental conditions toward an optimum result with respect to the observed response(s).

The problem is that the variations depicted in Fig. 1 are not independent: A sluggish substrate can often be brought to reaction by using a more aggressive reagent or by using forced reaction conditions. A change of solvent will alter the solvation of substrate and reagent(s) and this may modify the reactivity pattern. The yields of the desired product and the byproducts will usually respond differently to a change of the reaction conditions.

The interdependence of the possible variations makes it mandatory to use strategies which can take the joint influence of all factors simultaneously into account, i.e. to use multivariate methods. To be able to detect interaction effects it is necessary to design the experiments in such a way that the pertinent factors are varied simultaneously.

We note that the traditional method of studying experimental variables, viz. to examine one variable at a time (OVAT), while maintaining all the remaining variables at constant values is a notoriously poor strategy [5] which is bound to fail when there are interactions between the variables. The criticizm applies to variations in both the reaction space and the experimental space. It is also important to avoid situations in which two or more variables are changed in a correlated way over a series of experiments. In such experiments it will be impossible to discern which variable is responsible for an observed variation of the response.

The difficulties indicated can be overcome if experimental designs based upon statistical principles are used to vary the settings of the experimental factors.

In principle, it would be possible to predict the outcome of any synthetic reaction by using quantum mechanics and known physical and chemical models and through these derive how an optimum result should be obtained. The CAMEO program developed by Jorgensen [6] is an attempt in this direction. However, for a theoretical approach to be successful, the settings of *all* influencing variables have to be known. In many cases, except the most simple ones, this implies drastic approximations due to the complexity of the system. Predictions by such "theoretical" models will therefore be imprecise and will not be very useful for practical purposes. For this, it will be necessary to approach the problem from another direction, viz. to use experiments for establishing models for predictions and simulations.

3 Methods for Exploring the Experimental Space

3.1 Modelling

Although it is difficult to derive a practically useful analytical expression for a "theoretical" model, f, which relates the conditions of a synthetic reaction to the observed result, it is reasonable assume that such a functional dependence exists and that:

$$\text{Result} = f \text{ (experimental conditions)} \tag{1}$$

The outcome of a chemical reaction depends on the energetics of the reaction system. The nature of substrate, reagent(s), solvent and the detailed experimental conditions will define a potential energy surface for the reaction system. A reaction can be described as a path over this potential surface. A function describing this potential surface is ultimately derived from quantum mechanical principles. For describing macroscopic phenomena, it is reasonable to assume that such a function is smooth and several times differentiable. The function, f, which describes the observed outcome of a synthetic reaction will thus be derived from the potential energy function and we may assume that such functions are also smooth and differentiable provided that the variations in the experimental conditions are not too large. This makes it possible to obtain a local approximation of f by a Taylor expansion which will describe the general features of the function in a limited domain of variation. Local Taylor expansion models can be established through experiments and such experimentally derived models are excellent tools for studying synthetic procedures.

For optimizing a synthetic reaction with respect to the experimental conditions it is necessary to determine *how* the experimental variables exert their influence.

However, the "true" response, η, in an experiment cannot be observed directly. Instead we must use an experimentally estimated response, y, and like all experimental data, y is subject to an experimental error, e:

$$y = \eta + e \tag{2}$$

For instance, an isolated yield is always lower than the actual conversion due to losses during work-up.

The error term, e, contains both systematic and random errors.

Systematic errors will be present if the method used to determine y underestimates or overestimates η.

Random errors are always present and some sources are: small random fluctuations in the settings of the variables which result in that the same experiments never yield exactly the same result upon repeated runs; random variations of unknown factors which influence the result but which are not controlled in the experiment; random events during the measurement process, e.g. spikes in signals from detectors, temperature variations in recording spectra.

In the absence of systematic errors, the observed response, y, will be an unbiassed estimate of the true response, and the error term can be analyzed by statistical methods. In carefully executed synthesis experiments it is reasonable to assume that random errors occur independently of each other and that the observed variation of these random events are normally and independently distributed. A variation of the experimental conditions is considered to be significant if it produces a variation of the observed response outside the noise level given by the experimental error. Significant variations in this respect can then be analyzed by comparison to the error variation through known statistical distributions based on the normal distribution, e.g. the t distribution, the F distribution, and the χ^2 distribution.

The experimental conditions are defined by the settings of the experimental factors. Such factors will be denoted by x_i, where x is the setting of variable i.

A function which relates the observed response to the experimental factors can thus be written

$$y = f(x_1, \ldots x_k) + e \tag{3}$$

It is usual to scale the natural variable (e.g. temperature) by a linear transformation so that the variation of x_i is centred around a value of zero.

A Taylor expansion of f will have a form

$$y = \beta_0 + \sum \beta_i x_i + \sum\sum \beta_{ij} x_i x_j + \sum\sum \beta_{ii} x_i^2 + \ldots + R(x) + e \tag{4}$$

The Taylor approximation model is a polynomial in the experimental factors. The rest term, $R(x)$, becomes smaller and smaller the more polynomial terms are included in the model. $R(x)$ accounts for the variation which is not described by the polynomial terms and will thus contain the model error. A model is considered as satisfactory if the model error is significantly less than the experimental error.

Sufficiently good approximations are in most cases obtained by including up to second-degree terms in the model.

Such polynomial models are often called *response surface models* since they define a surface in the space spanned by $\{y, x_1, \ldots, x_k\}$.

The "true" model parameters $(\beta_i, \beta_{ij}, \ldots)$ are partial derivatives of the response function f and cannot be measured directly. It is, however, possible to otain estimates, b_i, b_{ij}, b_{ii}, of these parameters by multiple regression methods in which the polynomial model is fitted to known experimental results obtained by varying the settings of $x_{i'}$. These variations will then define an experimental design and are conveniently displayed as a *design matrix*, **D**, in which the rows describe the settings in the individual experiments and the columns describe the variations of the experimental variables over the series of experiments.

$$
\mathbf{D} =
\begin{bmatrix}
x_1 & x_2 & \ldots & x_k \\
x_{11} & x_{12} & \ldots & x_{k1} \\
x_{21} & x_{22} & \ldots & x_{2k} \\
\cdot & \cdot & & \cdot \\
\cdot & \cdot & & \cdot \\
\cdot & \cdot & & \cdot \\
x_{n1} & x_{n2} & \ldots & x_{nk}
\end{bmatrix}
\begin{array}{l}
\text{Exp no} \\
1 \\
2 \\
\cdot \\
\cdot \\
\cdot \\
n
\end{array}
\tag{5}
$$

The response surface models can be written in matrix notation as

$$
y = [1 x_1 x_2 \ldots x_k x_1 x_2 \ldots x_i x_j \ldots x_1^2 \ldots x_k^2] \boldsymbol{\beta} + e \tag{6}
$$

in which $\boldsymbol{\beta} = [\beta_0 \beta_1 \beta_2 \ldots \beta_k \beta_{12} \ldots \beta_{ij} \ldots \beta_{11} \ldots \beta_{kk}]'$ is the vector of "true" model parameters.

For the series of n experiments given by the design matrix **D** we thus obtain

$$
y_1 = [1 x_{11} x_{12} \ldots x_{1k} x_{11} x_{12} \ldots x_{1i} x_{1j} \ldots x_{11}^2 \ldots x_{1k}^2] \boldsymbol{\beta} + e_1
$$
$$
y_2 = [1 x_{21} x_{22} \ldots x_{2k} x_{21} x_{22} \ldots x_{2i} x_{2j} \ldots x_{21}^2 \ldots x_{2k}^2] \boldsymbol{\beta} + e_2
$$

$$
\cdot
$$
$$
\cdot \tag{7}
$$
$$
\cdot
$$

$$
y_n = [1 x_{n1} x_{n2} \ldots x_{nk} x_{n1} x_{n2} \ldots x_{ni} x_{nj} \ldots x_{n1}^2 \ldots x_{nk}^2] \boldsymbol{\beta} + e_n
$$

This is more conveniently expressed in matrix notation

$$
y = \mathbf{X}\boldsymbol{\beta} + e \tag{8}
$$

where $y = [y_1 y_2 \ldots y_n]'$ is the vector of observed responses; **X** is the *model matrix* obtained by appending a column of ones (corresponds to the constant β_0) and columns for cross-products and squares of the experimental variables; $e = [e_1 e_2 \ldots e_n]'$ is the vector of unknown experimental errors. A vector $b = [b_0 b_1 b_2 \ldots b_{12}$

... b_{ij} ... b_{11} ... $b_{kk}]'$ of least squares estimates of the model parameters is obtained by the computations given by

$$b = (X'X)^{-1} X'y \tag{9}$$

The matrix $(X'X)^{-1}$ is called the dispersion matrix and its properties are related to different quality criteria of the estimated parameters. Multiplication of $(X'X)^{-1}$ by an estimate of the experimental error variance gives the *variance-covariance matrix* of the regression in which the diagonal elements give the variances of the estimated parameters and hence, their standard errors. The off-diagonal elements give the covariances of the estimates parameters. Independent estimates can be obtained if the experiments are arranged so that $(X'X)^{-1}$ is a diagonal matrix. An experimental design for which this applies is called an *orthogonal design*. The columns of X are mutually orthogonal. The standard errors of the estimates are proportional to the square-roots of the eigenvalues of the dispersion matrix. The joint confidence region of the estimated parameters and hence, the overall precision of the predictions by the model is proportional to the square-root of the determinant $|(X'X)^{-1}|$. A maximum overall precision in the estimation of all model parameters will thus be obtained if this determinant is as small as possible. A design for which this is fulfilled is called a D-optimal design ("D" stands for "determinant"). As the dispersion matrix is obtained from the model matrix X which in turn is derived from the design matrix D, it is evident that the quality of the information obtained by the model is solely determined by the experimental design.

By a carefull spacing of the settings of the variables it is possible to obtain estimates of the parameters so that they describe the following different features of the fitted response surface model *and nothing else*:

The linear coefficients, b_i, describe the slope of the surface in the x_i direction.

The cross-product coefficients, b_{ij}, describe the twist of the surface over the plane spanned by x_i and x_j and thus accounts for an interaction effect between these variables.

The quadratic coefficients, b_{ii}, describe the curvature of the response surface in the x_i direction.

An appropriate experimental design will thus allow the influence of each individual experimental factor to be distinguished.

A thorough discussion on regression analysis has been given by Draper and Smith [7].

3.2 Two Common Problems: Screening and Optimization

For any synthetic procedure there are a large number of experimental variables which can be varied to define the detailed experimental conditions. It is not likely that all these variables will have a significant influence on the result but some of them will. The first problem is therefore to determine *which ones*. To this end, it will be necessary to run a screening experiment for sorting out the important experimental variables. This can be accomplished by fitting a low-degree polynomial response surface model which includes all variables considered, and then to use the model to trace the influence of the variables.

The next problem will be to determine how the significant variables should be adjusted to ensure an optimum result. This can be accomplished in different ways:

The method which yields the most accurate predictions of the optimum conditions is to use response surface modelling and to determine a model which also includes square terms to account for curvatures of the response surface. Analysis of the fitted model will then reveal how the experimental conditions should be adjusted to achieve an optimum result.

Sometimes it is not necessary to determine a response surface model tor locate the optimum conditions. Hill-climbing by direct search methods, e.g. search along the path of steepest ascent [8] or sequential simplex search [9], will lead to a point on the response surface near the optimum. The computations involved in these methods are rather trivial and do not require a computer and will for this reason not be discussed further in this chapter. Readers who require details of these direct search methods should consult Refs. [1, 8, 9].

Which type of design will be most efficient in a given situation is dependent on what type of information is sought. Other aspects will also be involved, e.g. how many individual experiments can we afford to run and how drastic are the approximations we can tolerate in the models. Generally speaking, the number of experiments will increase with the number of variables to study and with the degree of accuracy in the approximation of the response function.

3.3 Screening Experiments

3.3.1 Models for Screening

At the outset of the study of a new reaction very little is known with certainty. At this stage, we are not interested in very precise descriptions. It is sufficient to know whether or not the variables influence, what magnitude and direction these influences have, and whether there are important interaction effects.

Response models which are useful in this context are:
the linear models

$$y = \beta_0 + \sum \beta_i x_i + e \tag{10}$$

and the second-order interaction models

$$y = \beta_0 + \sum \beta_i x_i + \sum\sum \beta_{ij} x_i x_j + e \tag{11}$$

The linear models will approximate the response surface by a plane. The estimated linear coefficients, b_i, describe the slope of the plane and hence, the sensitivity of the response to a variation of the corresponding variable settings. A slope close to zero (of the same order as the experimental error) indicates that the variable does not have a significant influence, whereas a slope distinctly different from zero shows a significant influence.

11

In the second-order interaction model cross-product terms have also been included. A significant estimated cross-product coefficient, b_{ij} , shows that the influence of variable x_i is dependent on the settings of variable x_j, i.e. there is an interaction effect between these variables. Geometrically this corresponds to a twist of the response surface over the x_i, x_j plane.

3.3.2 Two-Level Designs

For obtaining estimates of the coefficients in linear and second-order interaction models it is sufficient to examine each variable at only two levels. These levels are usually selected to cover the interesting range of the variable settings. The models are most conveniently expressed in the coded variables x_i such that $(x_i = -1)$ and $(x_i = 1)$ correspond to the low $(-)$ level and the high $(+)$ level respectively. With such scaling, the numerical values of the estimated coefficients are a direct measure of the influence of the variation of the corresponding. In this context, classical two-level designs such as factorial and fractional factorial designs and Plackett-Burman designs are convenient.

An excellent introduction to experimental design has been given by Box, Hunter and Hunter [5]. A general treatment is found in the book by Cochran dan Cox [10]. Many introductory text-books have appeared in recent years, some examples of which are given in the reference list [11].

3.3.2.1 Factorial and Fractional Factorial Designs

A two-level factorial design contains all combinations of the settings of the variables. With k variables this gives a total of 2^k individual experimental runs. This type of design is convenient when the number of variables is limited, $k \leq 4$, but will rapidly give a prohibitively large number of experiments when the number of variables increases. Fortunately, to determine screening models it is possible to select small subsets (fractional factorial designs) of the possible combinations in complete factorial designs and yet obtain good estimates of the model parameters. The number of necessary individual experimental runs will be of the same order as the number of parameters to estimate. If only linear models are considered, it is possible to study three variables in four experiments (constant term, three linear coefficients); up to seven variables in eight runs; up to fifteen variables in sixteen runs, and so on. Such designs will be 2^{-p} (1/2, 1/4, 1/8 ... 1/2^p) fractions of the complete 2^k factorial design.

If also interaction effects are to be evaluated, second-order interaction models can be determined from a limited number of runs: four variables in $(3/4 \times 2^4)$ twelve runs (constant term, four linear coefficients, and six cross-product coefficients); five variables in sixteen runs (constant term, five linear coefficients, and ten cross-product coefficients).

A complete factorial design will span the experimental space very efficiently and contains all "corners" of the (hyper)cube defined by the $(-)$ and $(+)$ settings of the variable. A fractional factorial design will distribute the experimental point so that as much as possible of the variation of the experimental space is covered. One example is shown in Fig. 2 where four experiments in a fractional design

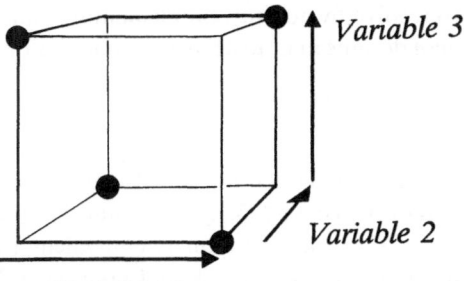

Variable 3

Variable 2

Variable 1

Fig. 2. Distribution of experimental point in a fractional factorial design in three variables

define a tetrahedron, which is the largest volume which can be spanned by four points in the three-dimensional variable space.

Fractional factorial designs are constructed from the model matrix X of a 2^{k-p} complete factorial design using the orthogonal columns of X to define the variable settings in 2^{k-p} experiments.

Of course, some information is lost by running fractions instead of complete factorial designs. In highly reduced fractional designs, the variations of the experimental variables x_i over the set of experiments will be identical to the variations of certain cross-products, $x_j x_k$ of other variables. From such experiments it is not possible to obtain individual estimates of linear and cross-product coefficients, they will be estimated as sums, $b_i = (\beta_i + \beta_{ik} + \dots \text{others})$. The "true" parameters are confounded. It is, however, possible to run complementary fractions so that any ambiguities in his respect can be resolved. Such complementary designs can be obtained by switching the signs of the variable settings in certain columns of the initial design matrix. The 2^{k-p} fractional factorial designs thus offer the great advantage of a sequential approach, i.e. a linear screening model is first adopted. If necessary, this model can be up-graded to a second-order interaction model by running one or more complementary fractions.

Thorough accounts of how to construct and evaluate factorial and fractional factorial designs are given in the literature [5], and we will no go into further details here. Computer programs which can assist in the lay-out of designs are listed at the end of this chapter.

Examples on the use of such designs in synthesis experiments are given below.

3.3.2.2 Plackett-Burman Designs

Another method for the construction of orthogonal screening design where the number, n, of experiments is a multiple of four ($n = 4, 8, 12, 16, 20, \dots$ up to 100) has been given by Plackett and Burman [12]. Such designs will also define fractions of complete factorial designs. The Placket-Burman designs used Hadamard matrices to define the settings of up to $(n - 1)$ variables in n runs. A Hadamard matrix, H, is a square matrix with the elements -1 or $+1$ and which has the properties that $H'H = n \times I_n$, i.e. the colums in H are mutually orthogonal.

The Plackett-Burman designs are convenient for fitting linear screening models when the number of variables is large and when it is desirable to keep the number of necessary runs to a minimum. One disadvantage is that the confounding patterns are complicated and quite cumbersome to resolve by adding complementary runs. For examples on the use of Plackett-Burman designs in industrial experimentation, see Ref. [11b].

3.3.2.3 D-Optimal Design

Sometimes, a reasonable response model can be suggested *a priori*. In such cases it is possible to use the D-optimality criterion, see Sect. 3.1., for establishing a suitable screening design. As $[(\mathbf{X}'\mathbf{X})^{-1}| = 1/|\mathbf{X}'\mathbf{X}|$ this criterion is equivalent to selecting a set of experiments which maximize the determinant $|\mathbf{X}'\mathbf{X}|$. In the general case it will be necessary to use a computer for establishing D-optimal designs. To use this criterion, the experimenter must specify a set of candidate experiments and the number of desired runs in the final design.

Several algorithms have been given by which experiments can be selected to yield a D-optimal design [13, 14]. A thorough treatment of the theory of optimal designs has been given by Fedorov. [14] D-optimal designs are very useful when the experimental domain is constrained so that certain combinations of the experimental factor settings cannot be used. Another situation when these designs are useful is when a minimum number of runs is highly desirable. A D-optimal design will select the best minimum subset for estimating the model parameters.

An example of a D-optimal design for a screening experiment in enamine synthesis is given in Sect. 3.4.4.

3.3.3 Identification of Significant Variables

Computerized routines for plotting make it possible to obtain graphical illustrations by projections of the response surface models and such plots are often sufficient for identifying influential variables. An example from the reduction of enamines is shown below.

When an estimate of the experimental error variance is known, it is possible to use statistical tests to evaluate the significance of the experimental variable. Analysis of variance can be used to partition the total sum of squares over the individual terms in the models and using the F distribution to determine which variations are significantly above the error noise level. It is also possible to use the t distribution to determine cofidence limits of the estimated model parameters and through these determine which ones have values which are significantly different from zero. Estimates of the experimental error variance can be obtained from the residual sum of squares after fitting the model to the experimentally observed results. These principles are illustrated below by an example on catalytic hydrogenation of furan. An independent measure of the experimental error variance can, of course, be obtained by replications of one or more experiments. For this purpose, it is strongly reommended that replicates of the experiments which corresponds to the average settings of the experimental variables ($x_i = 0$) be performed, since such experiments also allow checks on whether or not there is significant curvature of the response surface.

However, when the number of experiments is of the same order as the number of estimated model parameters, which is often the case when fractional designs have been used, no useful estimate of the experimental error variance is available.

The degrees of freedom of the error variance estimate are too low to permit clear-cut decisions on the significance by using the t or F distributions. In this situation it is possible to use another technique, viz. cumulative normal probability plots for assessing significant variables without having access to an estimate of the experimental error variance. This very useful technique was introduced by Daniel [15]. Orthogonal designs are balanced designs in the respect that each variable column in the model matrix contains an equal number of $(-)$ and $(+)$ signs. Assume that the experiments have been run in random order to eliminate time-dependent systematic errors. If we also assume that the experimental variables do not have any influence on the response, the response surface will be completely flat in the explored domain. Under these assumptions, an observed variation in the experimental results will then be nothing more than a manifestation of the experimental error. If a model is fitted to the observed experimental results, the estimated model parameters will be merely different balanced averaged summations of a normally distributed experimental error. The set of estimated model parameters will then be a random sample of normally distributed errors. We can therefore expect that the sample will also be normally distributed. A plot of the estimated model parameters will then define a straight line when plotted as a cumulative probability distribution on Normal probability graph paper. Estimated parameters which describe average summations of error terms are thereby easily identified. Estimated model parameters projected as outliers from this line indicate measures of something more than a random error and are likely indications of real influence by the corresponding variables. An example of this technique, furnished by the Willgerodt-Kindler reaction is given in Sect. 3.4.3.

3.4 Examples: Factorial Design, Fractional Factorial Design, D-Optimal Design

3.4.1 Enamine Reduction, 2^2 Factorial Design [16]

Enamines can be reduced to the corresponding saturated amines by treatment with formic acid. A very simple experimental procedure can be used in which formic acid is added to the neat enamine at such a rate that foaming due to evolution of carbon dioxide can be kept under control. The reduction of the morpholine enamine from camphor was studied in a two-level factorial design in order to determine whether or not an excess of formic acid should be used and at which temperature the reaction should be run.

Rolf Carlson and Åke Nordahl

The experimental design and yields of bornylamine obtained are shown in Table 1. The scaling of the variables: x_1, the amount of formic acid, and x_2, the reaction temperature, are given in a note to Table 1.

Table 1. Experimental design and yield in the enamine reduction

Exp no	Variables[a]		Yield
	x_1	x_2	y
1	-1	-1	80.4
2	1	-1	72.4
3	-1	1	94.4
4	1	1	90.6

[a] Coding of variables: x_i, Definition, [(-1)-level, ($+1$)-level]: x_1, Amount of formic acid/enamin (mol/mol), [1.0, 1.5]; x_2, Reaction temperature (°C), [25, 100].

The second-order interaction model obtained from the experiments was

$$y = 84.5 - 3.0x_1 + 8.1x_2 + 1.1x_1x_2 + e \tag{12}$$

From the model it is seen that the best yield if found when x_1 is at its low level (-1) which corresponds to a stoichiometric amount of formic acid and that an excess is detrimental to the yield (the coefficient is negative). The temperature x_2 is the most important factor and should be at its upper level. The experimental conditions established for this very simple procedure could be applied as a method for the reduction of a number of enamines [17].

A plot showing the shape of the fitted response surface model is shown in Fig. 3. Three-dimensional projections of response surface models make it easy to see how

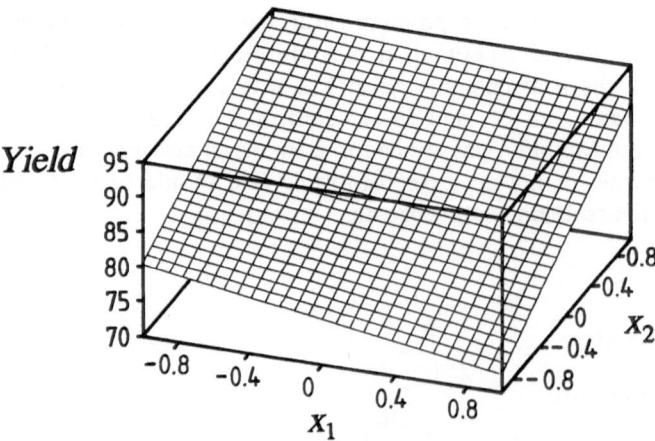

Fig. 3. Response surface showing the variation in yield in the reduction of camphor enamine

variations of the experimental conditions are transmitted to the variation of the response. Such projections are be available through several computer programs for response surface modelling. A list of such programs is gien in the Appendix.

3.4.2 Catalytic Hydrogenation of Furan, 2^4 Factorial Design [18]

The experiments were run with a view to determining how four experimental variables influence the yield in the catalytic hydrogenation of furan over a palladium catalyst. The variables were: x_1, the amount of palladium catalyst; x_2, the hydrogen pressure; x_3, the reaction temperature; and x_4, the stirring rate. The factorial design, 2^4, shown in Table 2 was used to fit a second-order intraction model to the yields obtained:

$$y = \beta_0 + \sum \beta_i x_i + \sum\sum \beta_{ij} x_i x_j + e \tag{13}$$

Table 2. Experimental design and yields obtained in the catalytic hydrogenation of furan

Exp no	Variables[a]				Yield
	x_1	x_2	x_3	x_4	y
1	−1	−1	−1	−1	77.5
2	1	−1	−1	−1	83.8
3	−1	1	−1	−1	87.8
4	1	1	−1	−1	92.9
5	−1	−1	1	−1	77.8
6	1	−1	1	−1	83.3
7	−1	1	1	−1	90.0
8	1	1	1	−1	94.1
9	−1	−1	−1	1	94.1
10	1	−1	−1	1	98.3
11	−1	1	−1	1	94.2
12	1	1	−1	1	98.3
13	−1	−1	1	1	97.0
14	1	−1	1	1	99.3
15	−1	1	1	1	98.0
16	1	1	1	1	100.0

[a] Coding of variables: x_i, Definition, [(−1)-level, (+1)-level]: x_1, Amount of catalyst/substrate (g/mol), [0.7, 1.0]; x_2, Hydrogen pressure (bar), [45, 55]; x_3, Reaction temperature (°C), [75, 100]; x_4, Stirring rate (rpm), [340, 475].

The model contains eleven parameters (constant term, four linear coefficients, six cross-product coefficients) and the design contains 16 experimental runs. With the assumption that interaction effects involving three or more factors have negligible influence on the yield, the residual sum of squares, $RSS = \sum e_i^2$, would then give an estimate of the experimental error variance, $s^2 = RSS/(16 - 11)$, with five degrees of freedom. the estimate of s^2 obtained in this way was used to compute 95% confidence limits of the estimated parameters.

The estimated parameters were

$$b_0 = 91.65 \pm 0.28 \qquad b_{12} = 0.10 \pm 0.28$$

$$b_1 = 2.10 \pm 0.28 \qquad b_{13} = -0.36 \pm 0.28$$

$$b_2 = 2.76 \pm 0.28 \qquad b_{14} = -0.52 \pm 0.28$$

$$b_3 = 0.79 \pm 0.28 \qquad b_{23} = 0.32 \pm 0.28$$

$$b_4 = 5.75 \pm 0.28 \qquad b_{24} = -2.54 \pm 0.28$$

$$b_{35} = 0.39 \pm 0.28$$

The confidence limits were computed as follows:

The total sum of squares, $SST = \sum y_i^2 = (77.5^2 + 83.8^2 + \ldots + 100.0^2)$ $= 135242.2$. The sum of squares due to regression, $SSR = b'X'Xb = 16 \times (b_0^2 + b_1^2 + \ldots + b_{34}^2) = 16 \times (91.65^2 + 2.10^2 + \ldots + 0.39^2) = 135241.2624$. The residual sum of squares, $RSS = SST - SSR = 135242.2 - 135241.2624 = 0.9676$. An estimate of the experimental error variance will thus be $s^2 = 0.9676/5 = 0.1852$. In a sixteen-run factorial design, the dispersion matrix is a diagonal matrix and $(X'X)^{-1} = 1/16 \times I$ and all parameters have an equal error variance $s^2/16$. The standard error of the estimated coefficients will therefore be $\sqrt{(0.1852/16)} \approx 0.108$. The critical t value for $\alpha = 5\%$ and five degrees of freedom is $t^{\mathrm{Crit.}} = 2.571$. The 95% confidence limits of the estimated parameters will thus be $\pm 2.751 \times 0.108 \approx 0.28$.

It is seen that all variables have significant linear influence, although the temperature variation, x_3, is not very important. One highly significant interaction effect, b_{24}, is found. This was an expected result. There is a compensation effect between hydrogen pressure, x_2, and the stirring rate, x_4. A low hydrogen pressure can be compensated by a high stirring rate, but this effect is less pronounced when

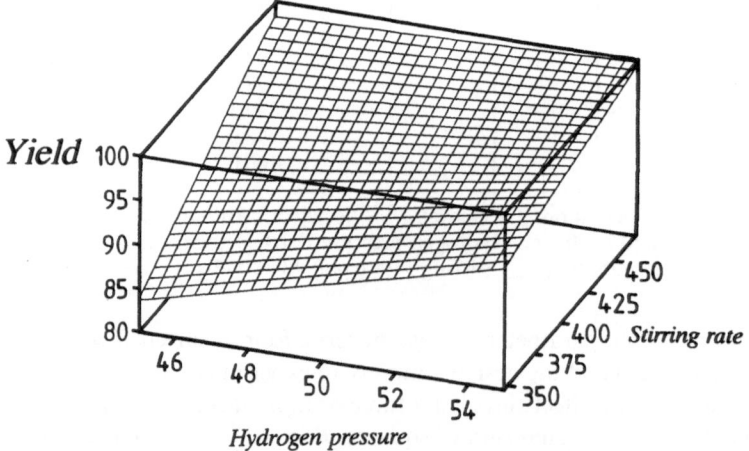

Fig. 4. Response surface showing the variation in yield when hydrogen pressure and stirring rate was varied in the catalytic hydrogenation of furan. The twist due to an interaction effect is clearly seen. The remaining variables were set to their upper levels

the hydrogen pressure increases. A projection of the response surface down to the space spanned by $\{y, x_2, x_4\}$ while setting the remaining variables at their high levels is shown in Fig. 4. The twist of the response surface plane due to the interaction effect is clearly seen.

3.4.3 Significant Variables in a Wilgerodt-Kindler Reaction, Fractional Factorial Design 2^{5-1}

The reaction of p-methyl acetophenone and morpholine with elemental sulfur in quinoline (solvent) was studied. [19] The objective was to determine which experimental variables have a significant influence on the yield.

Five variables were considered: x_1, the amount of sulfur; x_2, the amount of morpholine; x_3, the reaction temperature; x_4, the particle size of sulfur, and x_5, the stirring rate.

Table 3. Experimental design and yields obtained in the Willgerodt-Kindler screening experiment

Exp no	Variables[a]					Yield
	x_1	x_2	x_3	x_4	x_5	y
1	−1	−1	−1	−1	1	11.5
2	1	−1	−1	−1	−1	55.8
3	−1	1	−1	−1	−1	55.7
4	1	1	−1	−1	1	75.1
5	−1	−1	1	−1	−1	78.1
6	1	−1	1	−1	1	88.9
7	−1	1	1	−1	1	77.6
8	1	1	1	−1	−1	84.5
9	−1	−1	−1	1	−1	16.5
10	1	−1	−1	1	1	43.7
11	−1	1	−1	1	1	38.0
12	1	1	−1	1	−1	72.6
13	−1	−1	1	1	1	79.5
14	1	−1	1	1	−1	91.4
15	−1	1	1	1	−1	86.2
16	1	1	1	1	1	78.6

[a] Coding of variables: x_i, Definition, [(−1)-level, (+1)-level]: x_1, Amount of sulfur/ketone (mol/mol), [5, 11]; x_2, Amount of morpholine/ketone (mol/mol), [6, 11]; x_3, Reaction temperature (°C), [100, 140]; x_4, Particle size of sulfur (mesh), [240, 120]; x_5, Stirring rate (rpm), [300, 700].

As a generally accepted reaction mechanism has not yet been established it is not possible to exclude any interactions among the variables. A second-order interaction model was therefore indicated and a fractional factorial design, 2^{5-1}, was employed for obtaining estimates of the model parameters. The design is shown in Table 3. The scaling of the variables is given in a note to Table 3.

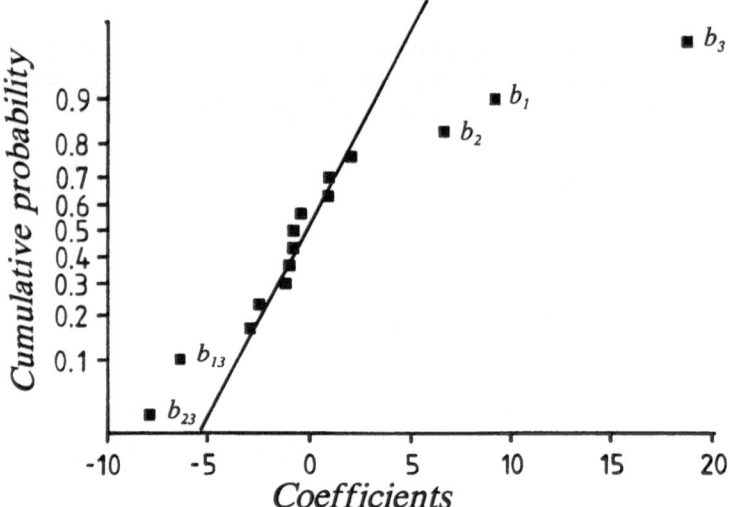

Fig. 5. Normal probability plot of estimated model parameters obtained in the screening experiment for the Willgerodt-Kindler reaction

A Normal probability plot of the estimated parameters is shown in Fig. 5. It is clearly seen that b_1, b_2, b_3, b_{13}, and b_{23} clearly fall outside the straight line that fits well to the remaining parameters. It was therefore concluded that the variables x_1, x_2, and x_3 as well as the interactions x_1x_3 and x_2x_3 have significant influences on the yield. All coefficients associated with variables x_4 and x_5 fall on the straight line and are best interpreted as measures of a normally distributed random experimental error. The variations explored for these variables have no significant influence on the yield. The conclusion was that for optimizing the procedure it would be sufficient to consider only the variables x_1, x_2, and x_3. This was fully confirmed in subsequent studies [19–22].

3.4.4 Enamine Synthesis over Molecular Sieves, D-Optimal Design [23]

A procedure for the formation of enamines by condensation of the parent ketone and secondary amines in the presence of molecular sieves was studied. The role of the molecular sieve is to trap the water formed. The enamine-forming reaction is acid catalyzed and both Brönsted acids and Lewis acids can be used.

The reaction between methyl isobutyl ketone and morpholine was studied with two commercially available molecular sieves (Type 3 Å and 5 Å). The procedure was described for other systems but the conclusions as to the amounts of amines to be used were contradictory [24]. These ambiguities needed to be clarified.

With a view to establishing suitable experimental conditions for preparative runs, the important experimental factors must be identified. To this end, a D-optimal design was used. The experimental variables considered are summarized in Table 4. It is seen that the variables describe both continuous and discrete variations. It is possible to use polynomial response function in such cases too, provided that the discrete variation is made to distinguish between only two alternatives. The model coefficients of the discrete variation is made to distinguish between only two alternatives. The model coefficients of the discrete variables describe the *systematic* variation of the response due to the alternatives.

Table 4. Variables and experimental domain in the D-optimal screening design of enamine synthesis over molecular sieves

Variables	Experimental domain	
	-1	1
x_1: The amount of morpholine/ketone (mol/mol)	1.0	3.0
x_2: The amount of solvent[a]/ketone (ml/mol)	200	800
x_3: The amount of molecular sieve/ketone (g/mol)	200	600
x_4: Type of molecular sieve	3 Å	5 Å
x_5: Type of acid catalyst[b]	SA	TFA
x_6: Amount of acid catalyst/ketone		
SA (g/mol)	50	200
TFA (ml/mol)	0.5	12.5

[a] Cyclohexane was used as solvent. [b] Two different types of acid catalysts were tested: *SA* was a solid Silica-Alumina, a Lewis acid-type of catalyst used in petroleum cracking; *TFA* was trifluoroactic acid, a proton acid.

Some of the variables could be expected also to have interaction effects:
— The amount of morpholine and the amount of solvent might influence the kinetics of the reaction and due to this, the yield evolution over time. To account for such effects, a term $\beta_{12}x_1x_2$ should be included in the model.
— There could be compensating effects between the amount of molecular sieve used and the type of molecular sieve used and to account for this, a term $\beta_{34}x_3x_4$ must be included.
— The same argument as above also applies to the type and amount of acid catalyst. For this reason, a term $\beta_{56}x_5x_6$ should be included.

It was assumed that a linear model augmented with the above-mentioned interaction terms would give an adequate description of the variation of the observed yield, y, i.e.

$$y = \beta_0 + \sum \beta_i x_i + \beta_{12}x_1x_2 + \beta_{34}x_3x_4 + \beta_{56}x_5x_6 + e \qquad (14)$$

21

The model contains ten parameter (constant term, six linear coefficients, and three cross-product coefficients. To estimate these coefficients, a design with twelve experiments was assumed to be sufficient. To span a maximum of variation the experimental design should include combinations of the variable setting at their high and low levels. With six variables at two levels, this gives a total of $2^6 = 64$ possible combinations. There are

$$64!/(52! \cdot 12!) \approx 3.28 \times 10^{12}$$

different ways to select twelve experiments from these 64 combinations and it is evident that a random selection runs the risk of being a very poor choice. For obaining a design which could give good estimates of the model parameters, the D-optimality criterion was used to select twelve experiments such that $|X'X|$ is maximized. For this purpose, the algorithm of exchange by Mitchell [13a] was used. In this procedure, an arbitrary set of twelve experimental runs (which may be drawn at random) is fed into the computer. By an iterative procedure, the initial set of experiments is modified by adding and deleting experiments so that the determinant $|X'X|$ is maximized. The design obtained in this case is given in Table 5 and the model matrix by this design afforded $|X'X| = 2.04 \times 10^{10}$. The estimated model parameters are summarized in Table 6.

An interpretation of the result is as follows. There are two effects which are significant beyond doubt: b_3, which is positive and implies that x_3, the amount of molecular sieve should be at least at its high level; b_4, which is also positive and shows that x_4 should be at its high level, i.e. a 5 Å type of molecular sieve should be used. The remaining variables do not influence the final yield. With short reaction time it is indicated that x_1 should be at its low level, $b_1^{28\,h} = -5.1$. This corresponds to a stoichiometric amount of morpholine. This effect can be understood as an effect of adsorption of morpholine on the molecular sieves. An excess of morpholine will block the acid catalytic sites on the surface of the molecular sieves, as well as the entrance to the channels in the zeolite crystals and this results in a lowered rate of enamine formation. This effect probably contributes

Table 5. D-optimal design in enamine synthesis, and yields, y_i obtained after "i" hours

Exp no	x_1	x_2	x_3	x_4	x_5	x_6	y_{28}	y_{52}	y_{76}	y_{98}
1	−1	−1	1	−1	1	1	58	66	77	76
2	−1	−1	−1	−1	1	−1	8	13	18	23
3	−1	−1	1	1	−1	1	60	61	65	67
4	−1	1	1	−1	−1	−1	40	60	69	74
5	1	−1	1	−1	1	−1	17	30	40	48
6	−1	1	1	1	1	−1	80	89	98	99
7	1	−1	−1	1	−1	−1	32	46	54	64
8	1	1	1	1	1	1	47	57	80	82
9	1	1	−1	−1	−1	−1	12	15	21	24
10	−1	1	−1	1	1	1	19	26	40	48
11	1	−1	−1	−1	−1	1	29	33	45	54
12	−1	1	−1	1	−1	1	46	40	57	60

Table 6. Estimated model parameters in the D-optimal screening of variables in the enamine synthesis

Parameters	Reaction time (h)			
	28	56	72	98
b_0	35.6	45.2	54.8	59.9
b_1	−5.1	−3.7	−0.8	−0.3
b_2	−1.5	1.1	2.9	1.4
b_3	14.4	15.7	16.3	15.1
b_4	7.9	8.3	8.7	8.3
b_5	−3.9	−4.8	−2.7	−2.4
b_6	−2.5	1.1	3.4	3.8
b_{12}	−3.2	−2.6	−4.4	−6.1
b_{34}	5.3	0.2	1.3	0.7
b_{56}	1.3	−0.4	0.4	1.2

to the slight increase of the interaction effect b_{12} over time. A more dilute solution will lower the amine adsorption and favour the reaction. For these reasons, a stoichiometric amount of amine is indicated. The suggested conditions for preparative purposes were therefore suggested to be:

x_1 should be at its low level which corresponds to a stoichiometric amount of amine;

x_2 is not too important. It is possible to use a small amount of solvent which is advantageous from an economic point of view;

x_3 a large amount of molecular sieve should be used;

x_4 molecular sieves of 5 Å type should be used;

x_5 the type of catalyst is not critical;

x_6 the amount of catalyst is sufficient at its low level.

In a final optimization of the procedure Silica-Alumina was chosen as catalyst. This was advantageous since it can be regenerated by the same procedure as the molecular sieves and the optimum mixture of the catalyst and the molecular sieve could be used for repeated runs after regeneration. A mixture of 750 g of molecular sieve and 250 g of the solid catalyst per mol of ketone afforded a quantitative yield after 70 h. The pure enamine could therefore be isolated after a simple work-up procedure involving only filtering of the reaction mixture and evaporation of the solvent.

3.5 Optimization

3.5.1 Response Surface Techniques
Response surface methodology was developed by Box and coworkers in the fifties [25]. Thorough treatments have been given in the literature [26] and the discussion here is given as a brief introduction.

At a true optimum, the response will have a maximum (or minimum) value. This means that there will be an extremum point on the response surface. To account for this, it is necessary to describe the curvature of the response surface in different directions. This can be accomplished by including also square terms, $\beta_{ii}x_i^2$, in the Taylor expansion approximation.

To reduce the number of necessary experiments, only significant variables should be considered at this stage. For obtaining estimates of the square coefficients, each variable must be investigated at more than two levels. Thus, three-level $[-1, 0, +1]$ factorial designs would be an obvious approach. However, the number of experiments, 3^k, increases rapidly with an increasing number, k, of variables, and such designs are convenient only for two-variable systems. With more variables it is possible to select small subsets of experiments from a three-level factorial design and thereby reduce the number of experiments. Box-Behnken designs [27a] and Hoke designs [27b] are examples of this. Another principle for reducing the number of experiments is to use "uniform shell designs" suggested by Doehlert [28]. In such designs, the experimental points (the settings of the variables) are evenly distributed on concentric (hyper)spheres in the experimental domain. However, the reduced three-level designs and the Doehlert designs have inferior statistical properties and a far better experimental lay-out can be obtained through the concept of rotatable designs which was introduced by Box and Hunter [29]. Among these, the central composite rotatable designs are excellent tools for the study of organic synthetic procedures since they permit a sequential approach to optimization.

3.5.2 Central Composite Designs

A central composite design consists of three parts:
— A factorial or fractional factorial design
— One or several experiments at the centre point (average settings of the experimental variables
— A set of experiments symmetrically displaced along the variable axes.

The distributions of experimental points for designs in two and three variables are shown in Fig. 6.

A response surface model can be used for predictions and simulations by entering values of variable settings in test points to the model. However, the model

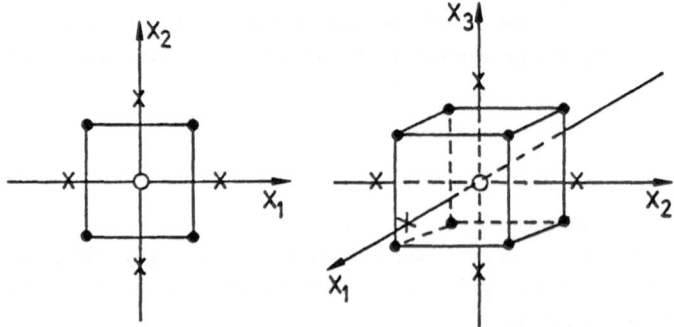

Fig. 6. Distribution of experimental points in central composite designs: ● factorial points, ○ centre point, × axial points

parameters are estimated from experimental results and hence, are not precisely known. They will have a probability distribution. The consequence of this is that there will be an error variance of the predictions. Rotatability means that this prediction error variance depends only on the Euclidian distance between the test point and the centre point. This is accomplished by a proper spacing of the experimental points, i.e. the experimental design used to establish the response surface model. A design which affords rotatability is called a rotatable design. It is possible to select a suitable number of experiments at the centre point so that the prediction error is fairly constant in the entire experimental domain, which an obvious advantage if the model is to be used for simulations.

3.5.3 A Sequential Approach is Possible

When a new reaction is explored, it is known beforehand whether it will be necessary to precisely locate the optimum conditions. If sometimes happens that the initial experiments reveal that the reaction does not afford the desired result. In such cases, the most probable decision is to turn to more promising projects. If, on the other hand, the screening experiments indicate the the reaction merits further studies, the next step would be to more precisely locate the optimum conditions.

An initial screening by a factorial or a fractional factorial design makes it possible to determine a second-order interaction model. In such models, the constant term, β_0, would correspond to the expected response when all (scaled) variables are set to their average (zero) values, i.e. the centre point of the experimental domain. If the screening models give an adequate description of the variation of the response, experiments run at the centre point should give a response which corresponds to b_0. Replication of the centre point experiment would give an estimate of the average response, $\bar{y}(0)$, at the centre point as well as an estimate of the experimental error variance. This makes it possible to determine whether or not an observed difference, $\bar{y}(0) - b_0$, is significantly outside the range of the experimental error variation. If this should be found be the case, the observed difference amounts to the sum of square coefficients, $\bar{y}(0) - b_0 = \sum \beta_{ii}$. Estimates of the individual square coefficients can then be obtained by running a complementary set of experiments in which the settings of the experimental variables are symmetrically displaced $\pm \alpha$ along the variable axes. Rotatability is achieved if $\alpha = (N_F)^{1/4}$, where N_F is the number of factorial points [29].

3.6 Modified TiCl$_4$ Method for Enamine Synthesis [30]

A standard method for enamine synthesis from aldehydes or ketones is to heat the carbonyl compound and the secondary amine in benzene or toluene and remove the water formed by azeotropic distillation. This method cannot, however, be used the preparation of enamines from methyl ketones which undergo self-condensation under these conditions. A procedure which overcomes these difficulties has been given by White and Weingarten [31]. The method employs anhydrous titanium tetrachloride as water scavenger. In the original procedure by White and Weingarten, titanium tetrachloride is added dropwise to a cooled solution of the carbonly compound and the amine, followed by prolonged stirring at room temperature. It was later found that considerably shortened reaction times could be achieved by adding the carbonyl compound to a performed complex between titanium tetrachloride and the amine [30]. By this modified procedure, almost quantitative yield could be obtained within a few minutes. It was, however, found that the relative amounts of $TiCl_4$/carbonyl compound and amine/carbonyl compound which should be used for achiving a rapid enamine formation were highly dependent on the structure of the carbonyl compound [30, 32]. In the example shown below, the synthesis of enamine from methyl isobutyl ketone and morpholine was studied by a central composite design.

$$TiCl_4 \; + \; HN\bigcirc O \longrightarrow [Complex] \longrightarrow$$

The experimental design and the yields obtained after 15 min are shown in Table 7. The quadratic response surface model fitted to the experiments in Table 7 was:

$$y = 93.46 + 2.22x_1 + 9.9x_2 + 3.42x_1x_2 - 4.37x_1^2 - 10.32x_2^2 + e$$
(15)

From the three-dimensional projection of the response surface in Fig. 7 it is seen that there is a maximum point on the surface within the explored domain. At this maximum point, the tangential plane to the surface has a zero slope in all

Table 7. Central composite design and yields obtained in the enamine synthesis by the $TiCl_4$ method

Exp no	Variables[a]		Yield
	x_1	x_2	y
1	−1	−1	73.4
2	1	−1	69.7
3	−1	1	88.7
4	1	1	98.7
5	−1.414	0	76.8
6	1.414	0	84.9
7	0	−1.414	56.6
8	0	1.414	81.3
9	0	0	96.3
10	0	0	96.4
11	0	0	87.5
12	0	0	96.1
13	0	0	90.5

[a] Coding of variables: x_i, Definition, [Levels: −1.414, −1, 0, +1, +1.414]: x_1, Amount of $TiCl_4$/ketone (mol/mol), [0.50, 0.57, 0.75, 0.93, 1.00]; x_2, Amount of morpholine/ketone (mol/mol), [3.0, 3.7, 5.5, 7.3, 8.0].

directions and the coordinates for this point can be determined by differentiation of the response surface model and solving the equation system obtained by setting the partial derivative with respect to x_1 and x_2 equal to zero. This affords

$$2.22 + 3.42x_2 - 8.74 = 0 \tag{16}$$

$$9.90 + 3.42x_1 - 20.64 = 0 \tag{17}$$

which gives the roots

$$x_1 = 0.4721 \approx 0.47 \tag{18}$$

$$x_2 = 0.5580 \approx 0.56 \tag{19}$$

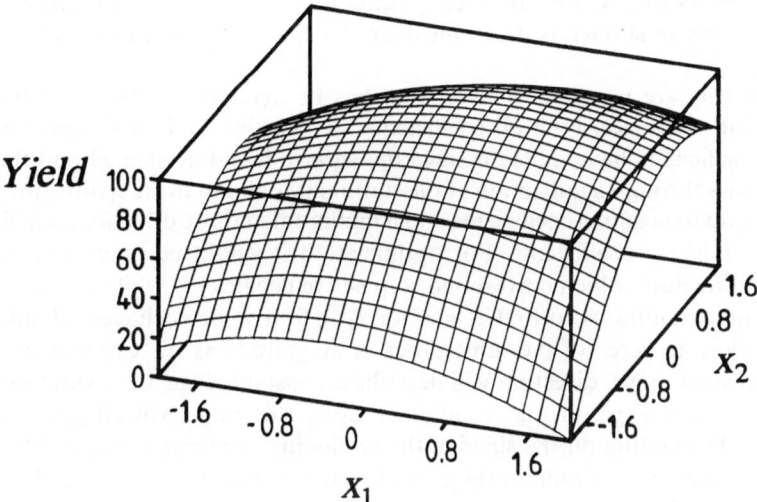

Fig. 7. Response surface showing the variation of the yield of enamine in the modified $TiCl_4$ procedure

The maximum point corresponds to the following molar ratios of the reagents: $TiCl_4$/ketone = 0.83 and morpholine/ketone = 6.5. Experiments carried out under these conditions afforded quantitative yields (98 ± 2%) within 15 min and this confirmed the optimum conditions.

3.7 Canonical Analysis of Response Surface Models [26]

In the example above, a maximum point was found within the explored domain. This is, however, not often encountered. Most frequently, the response surface is either monotonous in the explored domain or describes saddle-shaped surfaces or ridge systems. In such cases, it is not easy to comprehend the shape of the response surface from the algebraic expression of the model. A transformation to the canonical model will then be very helpful. Such a procedure constitutes a variable transformation from the coordinate system of the experimental variables $\{x_1, x_2, ..., x_k\}$ to another orthogonal system, $\{z_1, z_2, ..., z_k\}$ which has its origin at the stationary point on the response surface and which converts the mixed quadratic model to a polynomial which contains only quadratic terms. Geometrically, this transformation involves a translation of the origin of the original coordinate system to the stationary point and a rotation so that the axes of the new system are parallel with the main axes of the quadratic surface. This removes the linear and the cross-product terms and the model will be:

$$y = y_s + \lambda_1 z_1^2 + \lambda_2 z_2^2 + ... + \lambda_k z_k^2 + e \tag{20}$$

The constant term, y_s, is the predicted response at the stationary point on the response surface. As the square variables z_i^2 cannot be negative it is seen that the shape of the response surface is determined by the signs and magnitudes of the coefficients λ_i.

If all coefficients are negative, the response has its maximum value y_s at the stationary point. Analogously, if all coefficients are positive, y_s is the minimum value. If the coefficients have different signs the surface is saddle-shaped and the response will pass through a maximum at the stationary point in those z_i directions which correspond to negative coefficients and a minimum in those directions which correspond to positive coefficients. For optimization it would be interesting to explore the z_i directions which correspond to positive coefficients while adjusting lthe experimental conditions so that $z_j = 0$ for $\lambda_j < 0$. If some coefficient should have a value close to zero (of the same order of magnitude as the experimental error) the corresponding z_i direction will describe a constant ridge, i.e. a variation of the settings of the experimental conditions along this ridge will all give the same response. Depending on the signs of the remaining coefficients it can either be a maximum ridge or a minimum ridge. Such constant ridges occur when there is some functional dependence between the experimental variables and when a constant ridge is found this often can provide valuable information on the basic mechanisms of the system.

Response surface models are local Taylor expansion models which are valid only in the explored domain. It is often found that the stationary point on the response surface is remote from the explored domain and in the model may not describe any real phenomenon around the stationary point. Mathematically, a stationary point can be a maximum, a minimum, or a saddle point but it sometimes corresponds to unrealistic reponses (e.g. yield > 100%) or unattainable experimental conditions (e.g. negative concentrations of reactants). When the stationary point is outside the explored domain, the response surface is monotonous in the explored experimental domain and z_i directions which correspond to small coefficients will describe rising or falling ridges. Exploring such ridges offers a means for optimizing the response even if the response surface should be oddly shaped.

The mathematics involved in canonical analysis can be explained as follows.

In matrix notation, the fitted quadratic response surface model can be written

$$y = b_0 + \mathbf{x'b} + \mathbf{x'Bx} + e \tag{21}$$

where $\mathbf{x} = [x_1 \ldots x_k]'$ is the vector of the experimental variables; $\mathbf{b} = [b_1 \ldots b_k]'$ is the vector of linear model parameters; $\mathbf{B}$ is a symmetric quadratic matrix in which the diagonal elements are the quadratic coefficients, b_{ii}, and the off-diagonal elements are the cross-product coefficients divided by two, $b_{ij}/2 = b_{ji}/2$.

As $\mathbf{B}$ is a symmetric real matrix it can be factorized and expressed as the product of three matrices

$$\mathbf{B} = \mathbf{MLM'} \tag{22}$$

where $\mathbf{L}$ is a diagonal matrix in which the diagonal elements, λ_1, are the eigenvalues to $\mathbf{B}$, and $\mathbf{M}$ is an orthogonal matrix ($\mathbf{M'} = \mathbf{M}^{-1}$) in which the column vectors, $\mathbf{m}_i$, are the normalized eigenvectors to λ_i, i.e. $\mathbf{M} = [\mathbf{m}_1\mathbf{m}_2 \ldots \mathbf{m}_k]$. As $\mathbf{M}$ is an orthogonal matrix it is also true that

$$\mathbf{L} = \mathbf{M'BM} \tag{23}$$

Let $\mathbf{z} = [z_1 \ldots z_k]$ be the vector defined by

$$\mathbf{z} = \mathbf{M'x} \tag{24}$$

which is equivalent to

$$\mathbf{x} = \mathbf{Mz} \tag{25}$$

The matrix $\mathbf{M}$ describes a rotation of the $\{x_1 \ldots x_k\}$ system and the response surface expressed in z_i will thus be

$$y = b_0 + (\mathbf{Mz})' \, \mathbf{b} + (\mathbf{Mz})' \, \mathbf{B(Mz)} + e \tag{26}$$

i.e.

$$y = b_0 + \mathbf{z'M'b} + \mathbf{z'(M'BM)} \, \mathbf{z} + e \tag{27}$$

which can be written

$$y = b_0 + \mathbf{z'\theta} + \mathbf{z'Lz} + e \tag{28}$$

where $\mathbf{\theta} = \mathbf{M'b} = [\theta_1 \ldots \theta_k]'$ is the vector of linear coefficients in the general canonical model which can be written

$$y = b_0 + \theta_1 z_1 + \ldots + \theta_k z_k + \lambda_1 z_1^2 + \ldots + \lambda_k z_k^2 + e \tag{29}$$

29

The rotation removes the cross-product terms from the original response surface model. This form of the canonical model is very useful for exploring ridge systems. Such systems occur when some eigenvalue λ_i is small due to a feeble curvature of the surface in the corresponding z_i direction. The variation in this direction is therefore largely described by the corresponding linear coefficient θ_i.

The linear terms can be removed from the canonical model by shifting the origin of the $\{z_1 \ldots z_k\}$ system to the stationary point by the transformation

$$\mathbf{z} = \mathbf{M}'[\mathbf{x} - \mathbf{x}_S] \tag{30}$$

where $\mathbf{x}_S = [x_{1S} \ldots x_{kS}]'$ is the vector of the coordinates of the stationary point. This transformation yields

$$y = y_S + \lambda_1 z_1^2 + \ldots + \lambda_k z_k^2 + e$$

which is equivalent to Eq. (20).

The computations involved in canonical analysis cannot easily be carried out by hand and require a computer. Programs for response surface modelling and canonical analysis are given in the Appendix to this chapter.

3.7.1 Example: Synthesis of 2-Trimethylsilyloxy-1,3-Butadiene. Exploration of a Rising Ridge System by Canonical Analysis

It was found that methyl vinyl ketone could be converted into 2-trimethylsilyl-1,3-butadiene by reaction with chlorotrimethylsilane (TMSCl) in the presence of triethylamie (TEA) and lithium bromide in tetrahydrofuran solution [33].

The reaction uses readily available and cheap reagents and it was desired to determine under which experimental conditions a maximum yield, y, could be obtained. However, the stoichiometry of the reaction was not known and the first objective was therefore to determine what amounts of reagents with respect to the ketone substrate should be used. To this end, a response surface design was laid out in which the relative amounts of TMSCl, TEA, and LiBr were varied, see Table 8 [34].

The response surface·model stablished from the experiments in Table 8 was

$$y = 83.57 - 3.76x_1 - 0.96x_2 + 0.60x_3 + 1.71x_1x_2 - 0.41x_1x_3$$

$$- 0.46x_2x_3 - 0.63x_1^2 - 2.20x_2^2 - 0.81x_3^2 + e \tag{31}$$

Table 8. Design and yields obtained in the synthesis of 2-trimethyl-silyloxy-1,3-butadiene

Exp no	Variables[a]			Yield
	x_1	x_2	x_3	y
1	−1	−1	−1	82.9
2	1	−1	−1	72.1
3	−1	1	−1	82.5
4	1	1	−1	77.9
5	−1	−1	1	87.0
6	1	−1	1	73.9
7	−1	1	1	84.1
8	1	1	1	78.5
9	0	0	0	83.4
10	−1.75	0	0	87.1
11	0	−1.75	0	75.0
12	0	0	−1.75	80.8
13	1.75	0	0	76.2
14	0	1.75	0	78.7
15	0	0	1.75	81.0
16	0	0	0	83.6
17	0	0	0	84.2
18	0	0	0	82.7

[a] Coding of variables: x_i, Definition, [Levels: −1.75, −1, 0, +1, +1.75]: x_1, Amount of TEA/ketone (mol/mol), [1.56, 1.75, 2.00, 2.25, 2.44]; x_2, Amount of TMSCl/ketone (mol/mol), [0.97, 1.20, 1.50, 1.80, 2.03]; x_3, Amount of LiBr/ketone (mol/mol), [0.70, 1.00, 1.40, 1.80, 2.10].

By the transformation

$$x_1 = 0.8451z_1 + 0.3622z_2 - 0.3924z_3 - 8.6733 \tag{32}$$

$$x_2 = 0.3931z_1 + 0.0774z_2 + 0.9162z_3 - 3.4400 \tag{33}$$

$$x_3 = -0.8451z_1 + 0.3622z_2 - 0.3924z_3 - 8.6733 \tag{34}$$

the response surface model is transformed to

$$y = 99.23 - 0.15z_1^2 - 0.98z_2^2 - 2.59z_3^2 + e \tag{35}$$

The model indicates a maximum yield of 99.23% at the stationary point given by the vector $x_s = [-8.6733 \quad -3.4400 \quad 3.2737]'$. This is, however, far outside the explored domain and corresponds to unrealistic experimental conditions (e.g. negative amounts of TEA).

The coefficients for z_i in Eq. (32–34) are the elements of the eigenvector matrix **M** and transformation of **z** to **x** can be written

$$\mathbf{x} = \mathbf{Mz} + \mathbf{x_s} \tag{36}$$

If we set $\mathbf{x} = \mathbf{0}$ (the centre of the experimental domain) in the above expression we can compute $\mathbf{z}_0$ which is the vector $\mathbf{z}_0 = [z_{10}z_{20}z_{30}]'$ of the z_i coordinates of the centre point. This gives

$$\mathbf{z}_0 = -\mathbf{M}'\mathbf{x}_S = [9.8881 \ 0.3733 \ -0.4640]' \tag{37}$$

This shows that the z_1 axis passes the design centre at a distance of 9.8881, which is far outside the explored domain.

By the transformation

$$Z_1 = z_1 - 9.8881 \tag{38}$$

the origin of the canonical coordinate system is moved into the experimental domain and the Z_1 axis passes the design centre. Substituting z_1 for Z_1 in the canonical model yields after simplication

$$y = 84.29 - 3.02Z - 0.15Z_1^2 - 0.98z_2^2 - 2.59z_3^2 + e \tag{39}$$

This shows that the response surface in the explored domain is described by a maximum ridge along the Z_1 axis. The ridge is rising in the negative Z_1 direction. To increase the yield, the experimental variables should be adjusted so that z_2 and z_3 both have zero values. If this is fulfilled, an increased yield is to be expected if the experimental conditions are adjusted along the path which describes a displacement in the negative direction of the Z_1 axis. When the experimental conditions are adjusted according to this, the relative amounts of TEA:TMSCl: LiBr approach the relation 1:1:2. This is likely to reflect the critical stoichiometry of the reaction. For a maximum conversion a twofold excess of lithium bromide should be used.

The next step was to determine the influence of the reaction temperature and whether or not an excess of the reagents in the relative properties TEA:TMSCl: LiBr = 1:1:2 should be used. A second response surface model was therefore established. It was found that running the reaction at 40 °C and using 1.5 equivalents of TMS and TEA respectively, and 3 equivalents of lithium bromide afforded 96.7 and 98.2% yield in duplicate runs. These results were assumed to reflect closely the optimum conditions and were used in preparative runs.

The method developed by these experiments afforded excellent yields when it was applied to the synthesis of silyl enol ethers from a variety of ketones [33, 35].

4 Methods for Exploring the Reaction Space

4.1 The Reaction Space

In the sections above, we have discussed methods for exploring the experimental conditions. For such experiments, it is assumed that the reaction system (substrate, reagent(s), solvent) is defined and that variations with respect to discrete variations is limited to alternative choices. We shall now discuss what can be done when these requirements are not fulfilled and when the problem is first to determine a suitable reaction system which is worth optimizing with respect to the experimental conditions.

We define the reaction space as the union of all possible discrete variations of the nature of the substrate, of the reagent(s) and of the solvent. For almost any given reaction, the number of such possible combinations will be overwhelmingly large. The problem is therefore to select subsets of *representative* test systems which ensure a sufficient spread of the critical properties of the system to permit general conclusions to be drawn.

A problem is that discrete variations are prone to interaction effects, see Sect. 2.2. The ability to recognize such interaction effects is very important when new reactions are explored, otherwise the potential of the reaction for preparative purposes may be overlooked. It is therefore necessary to use multivariate strategies also for exploring the reaction space so that the joint influence of varying the substrate, reagent(s) and solvent can be evaluated.

To allow for a systematic search of the reaction space it is necessary to quantify the "axes". One problem is that discrete variation will change a number of molecular properties. A change of a substituent in the substrate will alter several properties, e.g. electron distribution, steric concestion, lipophilicity, hydrogen bond ability, etc. It is such "intrinsic" properties of the molecule which determine its chemical behaviour. By the concept of *principal properties* it is possible to obtain quantitative measures of the intrinsic properties. The principal properties can therefore be used to quantify the "axes" of the reaction space.

4.2 Principal Properties

When a molecule takes part in a reaction, it is properties at the molecular level which determine its chemical behaviour. Such intrinsic properties cannot be measured directly, however. What *can* be measured are macroscopic molecular properties which are likely to be manifestations of the intrinsic properties. It is therefore reasonable to assume that we can use macroscopic properties as probes on intrinsic properties. Through physical chemical models it is sometimes possible to relate macroscopic properties to intrinsic properties. For instance ^{13}C NMR shifts can be used to estimate electron densities on different carbon atoms in a molecule. It is reasonable to expect that macroscopic observable properties which depend on the *same* intrinsic property will be more or less correlated to each other. It is also likely that observed properties which depend on different intrinsic properties will not be strongly correlated. A few examples illustrate this: In a homologous series of compounds, the melting points and the boiling points are correlated. They depend on the strengths of intermolecular forces. To some extent such forces are due to van der Waals interactions, and hence, it is reasonable to assume a correlation also to the molar mass. Another example is furnished by the rather fuzzy concept "nucleophilicity". What is usually meant by this term is the ability to donate electron density to an electron-deficient site. A number of measurable properties are related to this intrinsic property, e.g. refractive index, basicity as measured by pK, ionization potential, HOMO-LUMO energies, $n - \pi^*$

transition energy, relative rates in standard reactions. It would however be quite foolish to assume that a variation of the measured value of the refractive index in a philosophical sense is responsible for an observed rate difference in a nucleophilic displacement reaction, i.e. we cannot assume a cause-effect relation between measures of macroscopic properties and chemical behavior in a reaction even if they are strongly correlated. They are related since they depend on the same underlying intrinsic property.

Organic chemistry is often described in terms of functional groups. Compounds are naturally grouped into classes which share a common functional group. Synthetic reactions usually involve transformations due to the presence of specific functionalities. Within a class of functionally related compounds, the compounds are in some respect similar to each other and it is reasonable to assume that a gradual variation of measured properties corresponds to a gradual variation of the intrinsic molecular properties.

Instrumental methods in chemistry make it possible to characterize any chemical compound by a very large number of different kind of measurements. Such data can be called *observables*. Examples are provided by: *Spectroscopy* (absorbtions in IR, NMR, UV, ESCA ...); *chromatography* (retentions in TLC, HPLC, GLC ...); *thermodynamics* (heat capacity, standard Gibbs energy of formation, heat of vaporization ...); *physical propery measures* (refractive index, boiling point, dielectric constant, dipole moment, solubility ...); *chemical properties* (protolytic constants, ionzation potential, lipophilicity ($\log P$) ...); *structural data* (bond lengths, bond angles, van der Waals radii ...); *empirical structural parameters* ($E_s, \sigma_I, \sigma^+, \ldots$).

For any given problem in which we have to consider the properties of a molecule, it is reasonable to assume that, at least, some of the observables will be related to the pertinent intrinsic properties. Analyzing the problem will make it possible to make certain *a priori* assumptions as to the relevance of the observables and those which we believe to be relevant to our problem will hereafter be called *descriptors*.

Assume that we have a set of n compounds which have all been characterized by a set of k different descriptors, q_i. Each compound will then yield a data vector $[q_1, q_2, \ldots, q_k]$. The data collected for the n compounds are most conveniently displayed as a table, a $(n \times k)$ data matrix $\mathbf{Q}$. In this matrix, the row vectors are the data vectors of the compounds. Such data matrices display two kinds of variation: Vertically, they show the between-compound variation of the descriptors; horizontally, they show the within-compound variation of the descriptors. These variations can be partitioned into two parts, *scores* and *loadings*, which can be analyzed separately. Mathematically, this is accomplished by a factorization of the data matrix. One method of doing this, which model the *systematic* variation, is to decompose $\mathbf{Q}$ into principal components.

4.3 Principal Component Analysis, PCA

The ideas behind principal component analysis were first described by Pearson [36] more than 90 years ago. The field of application at that time was in psychology. The method was further developed by Hotelling [37] in the thirties and is basically the same today. A similar but not identical method was developed by Thurnstone and is known as Factor analysis (FA) [38]. Sometimes FA and PCA are confused and incorrectly used as synonyms. Principal component analysis has been rediscovered many times in different disciplines and the nomenclature is bewildering. For instance, the method is known as singular value decomposition [39], Karhunen-Loeve expansion [40], eigenvector analysis or characteristic vector analysis [41]. In chemistry, the application of PCA was introduced by Malinowski [42]. Many applications in chemistry have emerged and more elaborate techniques based upon PCA and FA have been developed and include the SIMCA method for classification [43], evolving factor analysis (EFA) [44], heuristic evolving latent projections (HELP) [45], and target transformation factor analysis (TTFA) [46]. Instructive descriptions of the principles of PCA are found in Refs. [1, 47]. A review of the application of PCA in physical organic chemistry has been given Zalewski [48]. The use of PCA in organic synthesis is discussed in Ref. [1].

The basic ideas of principal component analysis are uncomplicated and easily understood from a geometric description. This is presented first. A brief account of the mathematics involved then follows.

4.3.1 Geometrical Description of PCA

If we let each descriptor variable, q_i, define a coordinate axis, the set of k descriptors defines a k-dimensional descriptor space. Although it is difficult to imagine more than three dimensions, geometrical concepts like points, lines, planes, distances will have the same meaning in k dimensions as they have in three dimensions. We can therefore use a three-dimensional space to illustrate the general case. In the k-dimensional descriptor space, each compound will correspond to a point with the coordinates along the descriptor axes equal to its measured properties. Similar compounds would have similar properties and they would be close to each other in the descriptor space. In such cases, they would be adequately described by the average value of the descriptors. However, if the compounds have different properties they would have a spread around the average point and the set of n compounds will describe a swarm of points with a distribution in the descriptor space. Compounds which are similar will be close to each other, while compounds which are dissimilar will be apart. Principal component analysis constitutes a projection of this swarm of points down to a space of fewer dimensions in such a way that as much as possible to the "shape" of the swarm is preserved in the projection. This means that compounds which are similar will be projected close to each other, while dissimilar objects will be projected at a certain distance to each other. The essence of this is that the *systematic* variation between the compounds will be portrayed by fewer variables than were present in the original descriptor space, viz. the axes which span the space of the projection. These axes are called the principal components. The principles are shown in Fig. 8.

The projection is made so that the first principal component vector, $\mathbf{p}_1$, describes the direction through the swarm of data points which shows the largest variation with respect to the distribution of the points in the space. If this vector is anchored at the average point, we can make perpendicular prjections of the data points on

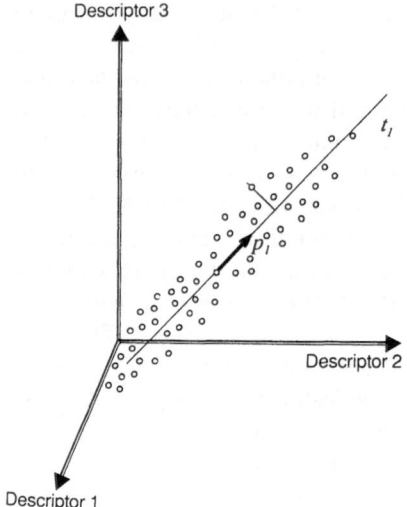

Fig. 8. Geometric illustration of principal component analysis

this vector. The distance of the projected points to the average point will thus define coordinates along the first principal component vector. These coordinates are called principal component *scores*. The score values measure how close the compounds are to each other with respect to the variation along the first principal component, and hence, the similarities between the compounds.

How much each descriptor contributes to the variation along the direction of the first component vector $\mathbf{p}_1$ is given by the cosine of the angle, φ_{1i}, between $\mathbf{p}_1$ and the descriptor axes $\mathbf{q}_i$. The vector $\mathbf{p}_1$ is defined by $\mathbf{p}_1 = [\cos \varphi_{11} \cos \varphi_{12} \ldots \cos \varphi_{1k}]'$. Normalizing this vector to unit length by dividing each element by $(\sum \cos^2 \varphi_{1i})^{0.5}$ gives $\mathbf{p}_1 = [p_{11} p_{21} \ldots p_{k1}]$. The p values are called *loadings*, and the vector is often called *loading vector*. The $\mathbf{p}_1$ vector will be more or less orthogonal to those descriptors which have only a feeble contribution to the variation, but tilted towards those descriptors which vary in a correlated manner over the set of compounds. As a correlated variation of measured properties is a manifestation of a variation of an underlying intrinsic property, it is seen that the principal component can be used to portray the intrinsic variation and hence, defines the "axes" of the reaction space. Those variations of measured properties which are described by principal component vectors will hereafter be called the *principal properties*.

The next step in the analysis is to determine whether there is systematic variation which was not accounted for by the first component and which could be described by a second component. The second component has a direction perpendicular to the first component and defines the direction through the swarm of points which describes the second next largest variation of the distribution of the data points. This constitutes a projection of the swarm of points to the plane spanned by the two first principal components. As the principal component vectors are orthogonal, they will portray different and independent principal properties.

This process is continued until all systematic variation has been exhausted and picked up by principal components and the variation which is left is nothing but a random error noise. How many components are significant can be determined by cross-validation, see below.

The variation of the principal properties in a set of compounds is quantified by the score values. This variation can be displayed by plotting the scores of different components against each other. Such score plots are very useful for selecting test compounds for experimental studies. Strategies for the selection of test system based upon principal properties are discussed in Sect. 4.6.

4.3.2 Mathematical Description of PCA

The raw data descriptor matrix Q usually contains descriptors of diffent kinds. The measured values may differ in magnitude (e.g. densities, interatomic distances, IR wave numbers) but also with respect to their range of variation. It is realized that the "shape" of the swarm of data points in the descriptor space is highly influenced by such factors. To avoid distorting the analysis by such factors, it will be necessary to standardize the axes of variation by a proper scaling of the original descriptors. A common procedure is to divide the values of each descriptor by its standard deviation over the set of compounds. This scaling (usually called outoscaling) gives each descriptor variable a variance equal to one over the set of compounds. For other types of scaling, see Wold et al. [47b]. Autoscaling also implies that we do not make any *a priori* assumptions as to the relevance of the various descriptor variables will be denoted by X.

The first step is to compute the averages of each descriptor variable. This yields the average vector, $\bar{x} = [\bar{x}_1 \bar{x}_2 \ldots \bar{x}_k]'$ which gives the average matrix, $\bar{X}$, after multiplication by the $(n \times 1)$ vector $1 = [11 \ldots 1]'$.

$$\bar{X} = 1\bar{x}' \tag{40}$$

The matrix X can then be written as a sum of the average matrix and a residual matrix E. The matrix E describes the variation around the mean. it is this variation which is described by the principal components.

$$X = \bar{X} + E \tag{41}$$

The variances and covariance of the descriptors are given by the matrix $(X - \bar{X})' (X - \bar{X})$, in which the diagonal elements are the variances of the variables and the off-diagonal elements are the covariances. When the data have been scaled to unit variance, this matrix is called the correlation matrix and the off-diagonal elements are correlation coefficients for the correlations between the variables, and the sum of the variances is equal to the number of variables.

As $(X - \bar{X})' (X - \bar{X})$ is a real symmetric matrix, it can be factorized into a diagonal matrix of eigenvalues, L, and matrices of the corresponding eigenvectors, P and P', cf canonical analysis (see Sect. 3.7)

$$(X - \bar{X})' (X - \bar{X}) = PLP' \tag{42}$$

The eigenvectors of $(\mathbf{X} - \bar{\mathbf{X}})'(\mathbf{X} - \bar{\mathbf{X}})$ are the principal component vectors $\mathbf{p}_i$ and the eigenvalues λ_i describe how much of the total variance is accounted for by $\mathbf{p}_i$.

The diagonal eigenvalue matrix $\mathbf{L}$ can be written as the product of two matrices $\mathbf{T}'\mathbf{T}$ in which the columns $\mathbf{t}_i$ are mutually orthogonal and have their scalar product $\mathbf{t}_i'\mathbf{t}_i = \lambda_i$. If we let $\mathbf{T}$ be a $(n \times k)$ matrix which obeys these criteria, then $\mathbf{T}'\mathbf{T} = \mathbf{L}$. We can therefore write

$$(\mathbf{X} - \bar{\mathbf{X}})'(\mathbf{X} - \bar{\mathbf{X}}) = \mathbf{PT}'\mathbf{TP}' \tag{43}$$

which is equivalent to

$$(\mathbf{X} - \bar{\mathbf{X}})'(\mathbf{X} - \bar{\mathbf{X}}) = (\mathbf{TP}')'(\mathbf{TP}') \tag{44}$$

which yields

$$(\mathbf{X} - \bar{\mathbf{X}}) = \mathbf{TP}' \tag{45}$$

It is seen that any $(n \times k)$ matrix can be factorized into a $(n \times k)$ score matrix $\mathbf{T}$ and an orthogonal eigenvector matrix. The elements of column vectors $\mathbf{t}_i$ in $\mathbf{T}$ are the score values of the compounds along the component vector $\mathbf{p}_i$. Actually, the score vectors are eigenvectors to $\mathbf{XX}'$.

A complete factorization corresponding to all non-zero eigenvalues is most conveniently achieved by the algorithm for singlar value decomposition given by Golub and Van Loan [39], which is more convenient than procedures involving diagonalization of the variance-covariance matrix.

It is, however, often found that only a few components are necessary to account for the systematic variation in the data. The remaining components have eigenvalues of the same order as the measurement error variance of the data. The significant components which describe the large and systematic variation of the data are called the *principal components*. For determining the principal components, it is not necessary to use procedures which involve computations of eigenvalues or diagonalization of the variance-covariance matrix. The NIPALS algorithm [49] makes it possible to use a sequential approach in which one PC dimension at a time is peeled off until the systematic variation has been described by the principal components. Another advantage of the NIPALS algorithm is that it can tolerate missing data in the descriptor matrix, an almost inevitable problem when descriptor data are compiled from literature sources.

The NIPALS algorithm uses a least squares fitting to determine the direction of the first component, i.e. so that the sum of squared deviations between the data points and their projections on the component vector is as small as possible. The first component is usually determined from the residual matrix obtained after subtraction of the average matrix $\bar{\mathbf{X}}$ from $\mathbf{X}$, i.e. $\mathbf{E} = \mathbf{X} - \bar{\mathbf{X}}$. The first component vector $\mathbf{p}_1$ is then used to compute the first score vector, $\mathbf{t}_1$.

$$\mathbf{t}_1 = \mathbf{Ep}_1 \tag{46}$$

The variation described by the first component is $t_1 p_1'$ which is a $(n \times k)$ matrix. If this is removed from E we obtain a new residual matrix which contains the variation not described by the average and the first component, i.e. $E = X - \bar{X} - t_1 p_1'$. This residual matrix is used for determining the second component, and so forth.

After the extraction of A components, we have

$$X = 1\bar{x}' + t_1 p_1' + t_2 p_2' + \dots + t_A p_A' + E \qquad (47)$$

which can be written

$$X = 1\bar{x}' + TP' + E \qquad (48)$$

The number of components that should be included in the model for describing the *systematic* variation of the data can be determined by cross-validation (CV). Details of cross validation in PCA have been given by Wold [50] and only a brief account of the principles is given here.

The PC model can be used for predictions. We can therefore determine the prediction error, $e_{ij}(\text{Pred.}) = x_{ij} - \hat{x}_{ij}$, to see how well an original datum x_{ij} is predicted, $\hat{x}_{ij}$. The NIPALS algorithm tolerates missing data in the X matrix. It is therefore possible to leave out data from X and compute a PC model from the truncated data set. The model can then be used to predict the left-out data and hence, to determine the prediction error. Assume that A components have been extracted from X. The residual matrix E_A then contains the deviations, e_{ij}, between the original data points and their projections on the A-dimensional (hyper)plane spanned by the A components. The sum of squared deviations, $\sum\sum e_{ij}^2$, is the residual sum of squares (RSS) efter fitting A components. If we proceed to extract another component $(A + 1)$ from E_A and in doing this, we leave out data in such a way that each datum in E_A is left out once and only once we can compute the prediction error sum of squares, PRESS $= \sum\sum e_{ij}(\text{Pred.})^2$ after $(A + 1)$ components. As the left-out data have not participated in the modelling, the errors of predictions are independent. The new component $(A + 1)$ is significant only if PRESS after $(A + 1)$ components is significantly less than RSS after A components. This can be evaluated by F statistics.

The cross-validation criterion is slightly conservative (rather too few than too many components are included) and ensures a maximum prediction ability of the PC model.

Principal component analysis can be achieved by many available computer programs. For chemical applications, where missing data are common, it will be necessary to use programs which employ the NIPALS algorithm. A list of programs is given in the Appendix at the end of this chapter.

4.4 Principal Properties and Organic Synthesis

Principal component analysis of property descriptors of classes of compounds reveals the principal properties. The variation of the principal properties between the compunds is measured by the scores. This variation is displayed by the score plots. A selection of test items which afford a spread of the corresponding points in the score plot is likely to correspond to a selection of test items which have spread in their intrinsic molecular properties. This offers an opportunity for selecting good test candidates for synthetic explorations. When selected items are different to each other they will increases the information content in each experiment since redundant experiments can be avoided. A selection by principal properties therefore renders experimentation more efficient.

To illustrate the principles, an example of the screening of suitable Lewis acid catalysts is first shown.

4.4.1 Example: Screening of Lewis acid Catalysts in Synthetic Procedures [51]

This study was made to check whether or not selections from score plots would permit suitable Lewis acid catalysts to be found.

A set of 28 Lewis acids were characterized by ten property descriptors compiled from the literature. The data are shown in Table 9.

Table 9. Lewis acids characterized by ten property descriptors

Acids	Descriptors									
	1	2	3	4	5	6	7	8	9	10
1 $AlCl_3$	2.1	704.2	628.8	110.47	91.84	2.26	102	15.8	12.01	—
2 BF_3	0	1137	1120.33	254.12	50.46	1.265	154	3.9	15.5	—
3 MoS_2	—	235.1	225.9	62.59	63.55	—	—	—	—	—
4 $SnCl_4$	0	511.3	440.1	258.6	165.3	2.43	7.6	2.87	—	−115
5 SO_2	1.63	320.5	—	237.6	—	1.4321	119	15.4	12.34	−18.2
6 $POCl_3$	2.4	519.1	520.8	222.46	138.78	1.95	122	13.3	11.89	−57.8
7 Me_3B	—	143.1	32.1	238.9	—	1.56	89	—	10.69	—
8 Me_3Al	—	136.4	99	209.41	155.6	—	61	2.9	9.76	—
9 Me_2SnCl_2	3.56	336.4	—	—	—	2.37	—	—	10.43	—
10 TiO_2	—	913.4	853.9	56.3	9.96	1.97	160	48	10.2	0
11 $ZnCl_2$	2.12	415.05	369.39	11.46	71.34	2.32	96	—	12.9	−65
12 $TiCl_3$	—	720.9	653.5	139.7	97.2	2.138	110	—	—	1110
13 $TiCl_4$	0	804.2	737.2	252.3	145.2	2.19	181	2.8	11.76	−54
14 VCl_4	0	576.8	503.27	242.44	—	2.03	92	—	—	1130
15 $CrCl_3$	—	395.2	356.1	114.5	71.1	2.12	91	—	9.97	6890
16 $MnCl_2$	—	481.3	440.3	118.2	72.9	2.32	98.8	—	11.02	14250
17 $FeCl_2$	—	341.79	302.3	117.95	76.65	2.38	95	—	10.34	14750
18 $FeCl_3$	1.28	404.6	398.3	146.4	128	2.32	81	—	—	—
19 $CoCl_2$	—	325.2	282.2	106.5	78.5	2.53	86	—	10.6	12660
20 $NiCl_2$	3.32	305.33	259.03	97.65	71.76	—	87	—	11.23	6145
21 $CuCl_2$	—	220.1	175.1	108.07	71.88	—	91.5	—	—	1080
22 $GaCl_3$	0.85	524.17	454.36	172	—	2.208	78.7	—	11.96	−63
23 $GeCl_4$	0	543.4	—	347.15	29.21	2.1	81	2.43	11.68	−72
24 $AsCl_3$	1.53	335.24	294.7	233.2	—	2.16	70	1.59	11.7	−79.9
25 BCl_3	0.61	427.2	387.4	206.3	106.7	1.75	109	0	11.62	−59.9
26 $SiCl_4$	0	601.54	569.32	328.6	145.17	2.019	95.3	2.4	12.06	−88.3
27 $SbCl_3$	3.9	381.75	324.1	186	104.87	2.325	74	33	10.75	−86.7
28 PCl_3	7.8	314.7	272.3	217.1	—	1.95	78.5	3.43	9.91	—

Principal component analysis yielded two significant components which described 54% of the total variance. The plots of the scores and loadings are shown in Fig. 7. From the loading plot is seen that the first component is largely described by the descriptors: 6 (mean bond length), 9 (ionization potential), and 10 (magnetic and diamagnetic susceptibility). The second component is described by: 5 (heat capacity), and 8 (dielectric constant). The descriptors 1 (dipole moment), 4 (standard entropy of formation), and 7 (mean bond energy) contributed to both components.

Three reactions, which were known from the literature to be catalyzed by Lewis acids were selected as test reactions. A, was the Reetz alkylation of silyl enol ethers with t-butyl chloride for which titanium tetrachloride is known to be useful [52]. B, was the Diels-Alder reaction between furan and acetylenedicarboxylic ester for which aluminium trichloride is a good catalyst [53]. C, was a Friedel-Crafts acylation for which aluminium trichloride is the preferred catalyst [54]. The reactions are summarized in Scheme 6.

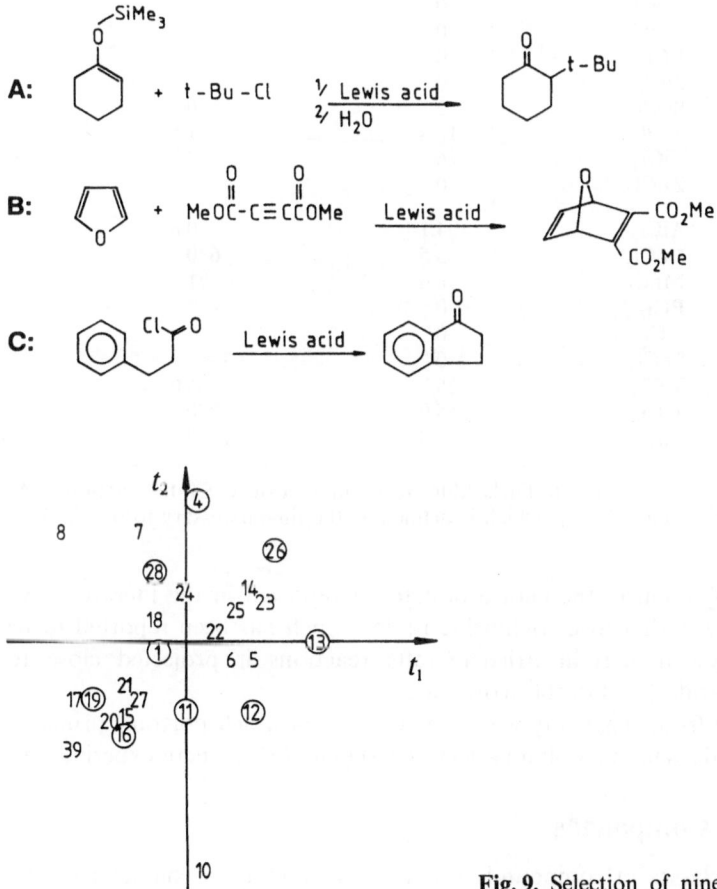

Fig. 9. Selection of nine Lewis acids from the score plot

41

Nine Lewis acids were selected from the score plot so that the selected points covered a variation, see Fig. 9. These acids were then checked in experimental runs with the three test reactions. The results are summarized in Table 10.

Table 10. Screening experiments with selected Lewis acids

Reaction[a]	Lewis Acid	Maximum yield (%)	Rate[b] t_{50} (min)
A	$AlCl_3$	39	1.9
	$CoCl_2$	0	–
	$MnCl_2$	0	–
	PCl_3	0	–
	$SiCl_4$	0	–
	$SnCl_4$	38	22.1
	$TiCl_3$	0	–
	$TiCl$	45	3.1
	$ZnCl_2$	44	378
B	$AlCl_3$	50	0.1
	$CoCl_2$	0	–
	$MnCl_2$	0	–
	PCl_3	0	–
	$SiCl_4$	0	–
	$SnCl_4$	2.5	20.2
	$TiCl_3$	13	1.5
	$TiCl_4$	26	7.0
	$ZnCl_2$	0	–
C	$AlCl_3$	94.1	0.8
	$CoCl_2$	3.5	630
	$MnCl_2$	6.6	71.2
	PCl_3	0	–
	$SiCl_4$	0	–
	$SnCl_4$	0	–
	$TiCl_3$	25.7	36.0
	$TiCl_4$	55.0	750
	$ZnCl_2$	5.0	1800

[a] A: Alkylation of silyl enol ether; B, Diels-Alder reaction, C, Friedel-Crafts reaction. [b] As a rough estimate of the kinetics, t_{50}, which is defined as the time necessary to obtain 50% of the final yield was used.

The results fully confirm the choice of catalyst reported in the literature. An interesting point was that iron trichloride (#18) which has been reported to be a superior catalyst in certain Friedel-Crafts reactions is projected close to aluminium trichloride (#1) in the score plot.

The conclusion from this study was that score plots which portrary principal properties were indeed useful tools for selecting test items for synthetic experiments.

4.5 Classes of Compounds

Up to now, the classes of compounds given below have been characterized by their principal properties for use in synthesis.

4.5.1 Organic Solvents [55]

The first example of principal properties for synthetic experiments was PC analysis of a set of 82 solvents, characterized by eight property descriptors [55]. The analysis was carried out with a view to finding rational principles for selecting test solvents for new reactions. Principal properties of solvents have now been determined from an augmented data set of 103 solvents characterized by nine descriptors [56]. A score plot is shown in Sect. 5.3.2 in the example on the Fischer indole synthesis.

Several PCA studies on solvents have been reported [57]. However, the objectives in these studies have been either to make classifications or to elucidate physical-chemical relations, and not explicitly to assist in synthetic experimentation. A discussion on multivariate studies on solvents is given in the book by Reichardt [58].

4.5.2 Lewis Acids [51]

After the initial study shown above, the stuy was augmented to include 116 Lewis acids and 20 descriptors [51]. Two significant components (cross-validation) accounted for 65% of the total variance. A score plot is shown in Sect. 5.3.2.

Selection of Lewis acids by their principal properties for synthetic exploration has been reported for syntheses of benzamides [59], of silyloxydienes [33], and of indoles [60].

4.5.3 Aldehydes and Ketones [61]

The most important single functional group in chemistry is the carbonyl group. The majority of synthetic reactions utilizes the presence of a carbonyl moiety for achieving the desired transformation. A number of factors will determine the reactivity of a carbonyl compound, e.g. thermodynamic and kinetic acidities of α hydrogens, stereoelectronic properties, steric hindrance for attack on the carbonyl group, solvent interactions (dipole-dipole, hydrogen bonding, Lewis acid-base interactions). A large number of elegant studies on the chemistry of the carbonyl group has been made, for examples, see Ref. [62]. Unfortunately, a serious problem when data are compiled from the literature is that many good descriptors are only available for a limited number of compounds. For practical utility in synthesis, principal properties must be determined from data sets of a sufficiently large number of compounds to cover the range of interest of the properties of potential test candidates. This limits the choice of descriptors. For characterizing aldehydes and ketones, seven common descriptors were compiled for commercially available compounds comprising 113 aldehydes, and 79 ketones. The descriptors were: molar mass, melting point, boiling point, density, refractive index, wave number of IR carbolyl band, molar volume. A study of a subset for which ^{13}C NMR shift of the carbonyl carbon was also available did not change the result in any significant way to what was obtained with the seen descriptors only [61].

Principal component analysis of the aldehydes and the ketones, respectively, afforded two significant components which accounted for 78% (aldehydes) and 88% (ketones) of the total variance. A score plot of the ketones is shown in the example of the Fischer indole synthesis given in Sect. 5.3.2. For a score plot of the aldehydes, see [61].

4.5.4 Amines [63]

A set of 126 primary, secondary, and tertiary amines characterized by seven property descriptors afforded two significant principal components which described 85% of the total variance. For details, see [63]. A preliminary study of 29 amines was used for the selection of co-substrates in studies on the Willgerodt-Kindler reaction [21].

4.5.5 Other Examples of the Use of Principal Properties

Characterization by principal properties has been reported for classes of compounds in applications other than organic synthesis: *Aminoacids*, where principal properties have been used for quantitative structure-activity relations (QSAR) of peptides [64]. *Environmentally hazardous chemicals*, for toxicity studies on homogeneous subgroups [65]. *Eluents for chromatography*, where principal properties of solvent mixtures have been used for optimization of chromatographic separations in HPLC and TLC [66].

4.6 Strategies for Selection

The systematic variation displayed by the score plots can be used in various ways for designing test sets for experimental studies. Which kind of design should be used will depend on the problem to be treated.

4.6.1 Use a Maximum Spread Design

Selection of test items so that they are as dissimilar a possible to each other is accomplished by choosing items which are projected far from each other and on the periphery of the plot. Such designs are useful for answering the question of whether or not the properties have an influence. A maximum spread design wa used to determine whether modification of the solvent would increase the *endo/exo* stereoselectivity in the reduction of an enamine from camphor. The answer was negative [17a].

Selections which afford a maximum spread also provide a good initial screening design for totally new reactions. Poor candidates can be eliminated early and the study can be continued with more promising candidates, see Sect. 4.6.2.

A maximum spread design can easily be augmented to a univorm spread design, see Sect. 4.6.3, and therebyl permits a sequential approach to the study of discrete variations. A small design which affords a large spread in the properties of the selected items also makes it possible to establish good initial PLS models, see Sect. 5.2.

4.6.2 Explore the Vicinity of a Promising Candidate

It is sometimes fairly well known which type of reagent or solvent should be used. To further improve the result, it might be profitable to investigate other candidates with similar properties. For this purpose, test items which are projected close to the "good" candidate should be chosen. We have found this principle to be useful for the selection of solvents for fractional crystallization.

4.6.3 Use a Uniform Spread Design

When the objective is either to make a carful screening for finding a good candidate for future development, or to study whether a gradual change in the performance of the reaction could be traced to the properties of the reaction system, a design which affords a selection of test items which are uniformly spread in the score plot should be employed. An example of this principle is given in the study of the Fisher indole reaction below.

4.6.4 Use a Statistical Design to Explore Several Dimensions of the Reaction Space

Often, it is of interest to study the joint influence of varying the substrate and/or the reagent(s) and/or the solvent. This implies that several dimensions of the experimental space should be explored. As interaction effects are to be expected, it is necessary to use a design in which all principal properties are varied simultaneously. This can be accomplished by factorial designs or fractional factorial designs in the principal properties.

A factorial design can be obtained as follows: For each dimension of the reaction space, use the score plot to select test candidates which afford a desired spread in the principal properties. Assume that: A substrates, B reagents, and C solvents have been selected. The design is to treat each substrate, with each reagent in each solvent. This yields a total of $A*B*C$ different combinations and is actually a multi-level factorial design. However, this usually gives a large number of test systems, see the example on the Fischer indole synthesis in Sect. 5.3.2 where a design with 600 systems was obtained in this way. Other examples which have been presented in the literature are the synthesis of benzamide by the direct reaction between benzoic acid and amines in the presence of a Lewis acid [59] where it was found that the proper choice of Lewis acid was dependent on the nature of the amine; and the synthesis of silyl enol ehters where an initial screening based upon these principles made it possible to find an optimal combination of reagent and solvent [33].

Complete multi-level factorial designs would usually yield too large a number of test systems for a first approach to new reaction systems. It is possible to reduce the number of test systems and yet achieve a selection which covers a large part of the entire reaction space. This can be achieved by a selection made from a two-level fractional factorial design. The principles are illustrated by an example provided by the Willgerodt-Kindler reaction.

4.7 Example: Fractional Factorial Design for Exploring the Reaction Space [22]

The Willgerodt-Kindler reaction was studied for a series of *para* substituted acetophenones with different amines in different solvents.

45

Substrate   Amine

The reaction system is characterized by the variation of the substrates, of the amine, and of the solvent. With the assumption that the variation of the substrate would be described by the properties of the *para* substituent, a principal component analysis of substituent parameters could be used for the selection of the substrates. Such a study was available [67] when this study was initiated. A two-component model described the variation in substituent propperties. Two-component models were also found for the amines and for the solvents. Hence, the reaction space is six-dimensional with respect to the principal properties. If two test items are selected for each principal propery axis so that these items span a large variation, this would define a two-level variation of each principal property and complete two-level factorial design in six variables would cover the entire reaction space. However, this would yield $2^6 = 64$ different system to test. By a fractional factorial design, it is possible to select subsets of these systems so that the subsets cover a much as possible of the reaction space. To span the range of variation of six variables an eight-run 2^{6-3} fractional factorial design would be sufficient. A design matrix is shown in Table 11.

Table 11. Fractional factorial design 2^{6-3} for screening of reaction systems

System no	Variables					
	1	2	3	4	5	6
1	−	−	−	+	+	+
2	+	−	−	−	−	+
3	−	+	−	−	+	−
4	+	+	−	+	−	−
5	−	−	+	+	−	−
6	+	−	+	−	+	−
7	−	+	+	−	−	+
8	+	+	+	+	+	+

The selection of test items by such a design can be accomplished as follows: For each constituent of the reaction system, two principal property axes should be considered. The columns of a two-level fractional factorial design matrix contain an equal number of minus and plus signs. If we let the columns pairwise define the selection of test systems, four combinations of signs are possible: [(−), (−)], [(−), (+)], [(+), (−)], and [(+), (+)]. These combinations of signs correspond to different quadrants in the score plots. Hence we can use the sign combinations of two columns to define from which quadrant in the score plot a test item should

Table 12. Assignments of design settings to selected test items

Assignment	Substituent	Amine	Solvent
− −	Cl-	Isopropylamine	Benzene
+ −	H-	Morpholine	Ethanol
− +	MeO-	Diethylamin	Quinoline
+ +	PhO-	Dipentylamine	Triethylene glycol (TEG)

be selected. Table 12 shows the constituents of the Willgerodt-Kindler system selected according to this criterion.

If we let columns in the design matrix define the constituents as follows: **1, 2** define the substrate, columns **3, 4** define the amine co-substrate, and columns **5, 6** define the solvent, the first row in the design matrix in Table 11 would thus correspond to a selection of a substrate projected in the $[(-), (-)]$ quadrant, an amine from the $[(-), (+)]$ quadrant, and a solvent from the $[(+), (+)]$ quadrant. The other rows define other combinations. The test items selected accordingly are shown in Table 13. To permit fair comparisons as to the performance of the reaction, it is necessary to adjust the experimental conditions for each system to yield an optimum result. The danger of using "standardized conditions" has been emphasized [1] and the arguments against such a technique are not repeated here. The conditions which afforded a maximum yield were determined by response surface techniques and these results are also shown in Table 13.

It is realized that it would be quite cumbersome to carry out elaborate optimizations for every new reaction system considered. Fortunately, this is not necessary since it is possible to use the technique of PLS modelling, described below, for determining quantitative relations between the properties of a reaction system and the performance of the system. The PLS model can then be used to

Table 13. Selected test systems and optimized experimental conditions

No	Test system			Optimum conditions[a]			Yield[b]
	Subst.	Amine	Solvent	S	N	T	(%)
1	Cl-	Et₂NH	TEG	8.4	5.3	123	89
2	H-	i-PrNH₂	Quinoline	10.3	4.8	133	89
3	MeO-	i-PrNH₂	EtOH	3.8	6.6	80[c]	91
4	PhO-	Et₂NH	Benzene	9.6	5.8	80[c]	85
5	Cl-	Pe₂NH	Benzene	13.6	8.5	80[c]	68
6	H-	Morpholine	EtOH	3.7	13.4	80[c]	86
7	MeO-	Morpholine	Quinoline	9.3	8.9	130	100
8	PhO-	Pe₂NH	TEG	13.0	8.3	118	73

[a] S, amount of sulfur/ketone (mol/mol); N, amount of amine/ketone (mol/mol); T, reaction temperature (°C). [b] Determined by gas chromatography (internal standard technique). [c] The reaction was carried out under reflux.

predict the performance of new, yet untested systems. A PLS model can be established from calibration sets which contain only a small number of systems, provided that these systems span the interesting range of variation.

If the Willgerodt-Kindler reaction were a recently discovered reaction and its scope had been tested according to the design discussed here, the results shown in Table 12 would have given the conclusions that the reaction seems to have a broad scope. Of course, it would have been impossible to draw any definite conclusions from such a small test set. An advantage of fractional factorial designs is that it is always possible to append complementary fractions so that an initial small design can be augmented to afford a better coverage of the experimental space. An important point is that even a small design permits PLS modelling, and complementary runs may serve two purpose; they can be used both to validate the model, and to upgrade the model. This can be repeated until a good description has been achieved.

Fractional factorial designs in the principal properties thus permit a sequential approach to the study of discrete variation. A sequential approach is beneficial, since it is not known beforehand whether a new reaction would yield results which are promising enough to motivate an elaborate study.

5 Quantitative Relations Between the Response Space, the Experimental Space, and the Reaction Space

In the above sections it was discussed how different perturbations which can be made when a synthetic reaction is run could be regarded as variations in two different spaces which span different kinds of possible variations of a reaction "theme". As there are often more than one response of interest, it is also necessary to consider the response space. In this section is discussed how these spaces can be connected by quantitatie models so that it will be possible to evaluate how perturbations in one space are related to variations in another space.

The general tool for this is PLS modelling. Before going into details of this method, we shall see how methods already described, screening designs and principal component analysis, can be combined and used to solve a common problem, viz. the evaluation of screening experiments when there are several response variables.

5.1 Evaluation of Multiresponse Screening Experiments

It is a common experience that an attempted synthetic transformation does not proceed cleanly, and that, in addition to the desired product, by-products and unreacted starting material are present in the final reaction mixture.

An early and important step will therefore be to identify which experimental factors are responsible for the gross and systematic variation of the *distribution* of the products in the reaction mixture. For this, a screening design which accommodates all pertinent experimental variables is appropriate. The obvious responses are the yields of the various products as well as the amounts of unreacted starting material(s) in the different runs. In experiments involving experimental variables, it is reasonable to assume cause-effect relations between the settings of the variables and the observed result. To account for this, response surface models obtained by multiple regressions can be used.

When there are only a few responses to consider, it is manageable to analyze each response variable separately, and then take all the results thus obtained together and make a joint evaluation. However, with more than a handful of responses such an approach will be quite difficult.

In an experimental design, the settings of the variables are given by the row vector $\mathbf{x} = [x_1 \ldots x_k]$ in the design matrix. When there are several responses, the result will define a response vector, $\mathbf{y} = [y_1 \ldots y_m]$. For the whole set of experiments these vectors define a *response matrix*, $\mathbf{Y}$.

If the products in the final reaction mixture in a synthetic experiment are formed by different and independent reaction mechanisms which operate during the course of the reaction, it is likely that these reaction mechanisms respond differently to the settings of the experimental variables. It is also likeyl that yields of products formed by different mechanisms vary independently of each other. On the other hand, yields of products which depend on the same mechanism will probably vary in a correlated way over the set of experiments. A principal component analysis of the response matrix would reveal such features. The method of analyzing multiresponse matrices by eigenvector decomposition was introduced by Box et al. [68] as a safeguard against erroneous regression models due to linear dependencies between responses. In a principal component model of the response matrix, responses which depend on different mechanisms will be portrayed by different components, while responses which depend on the same mechanisms will be described by the same component. The gross and systematic variations between the experimental runs will thus be portrayed by the score vectors. The influence of the experimental variables on this variation can be evaluated after fitting linear or second-order interaction screening models to the scores. The score value is a linear combination of the original responses. With the assumptions that the experimental errors of the original responses are normally and independently distributed, the scores will also have a normally distributed error. Significant experimental variables can therefore be identified from normal probability plots of the estimated parameters of the screening model [69]. An example to illustrate the principles is provided by the synthesis of substituted pyridines is used. Data have been taken from Diallo et al. [70].

5.1.1 Example: Evaluation of a Screening Experiment for the Synthesis of Pyridines

When 2-methyl-2-pentene is treated with acetyl chloride in the presence of a Lewis acid, a mixture of substituted pyrylium salts is formed. The pyrylium salts are transformed into the corresponding pyridines by threatment with aqueous ammonia, see Scheme 8.

Four experimental variables were assumed to influence the result and were studied by a two-level factorial design, see Table 14. In the original paper [70], the authors analyzed five derived responses which were calculated from the observed yields. The derived responses were analysed one at a time.

Principal component analysis of the response matrix afforded one significant component (cross-validation) which described 82% of total variance in Y. As all responses are of the same kind (percentage yield) the data were not autoscaled prior to analysis. The response y_{11} was deleted as it did not vary. The scores and loadings are also given in Table 14. The score values were used to fit a second-order

Table 14. Experimental variables and responses in the synthesis of pyridines

Entry	Experimental variables[a]				Entry	Experimental variables[a]			
	x_1	x_2	x_3	x_4		x_1	x_2	x_3	x_4
1	−	−	−	−	9	−	−	−	+
2	+	−	−	−	10	+	−	−	+
3	−	+	−	−	11	−	+	−	+
4	+	+	−	−	12	+	+	−	+
5	−	−	+	−	13	−	−	+	+
6	+	−	+	−	14	+	−	+	+
7	−	+	+	−	15	−	+	+	+
8	+	+	+	−	16	+	+	+	+

[a] Coding of variables: x_i, Definition [(−)-level, (+)-level]: x_1, Acetylchloride/Lewis acid ratio/(mol/mol), [0.8, 1.7]; x_2, Acetylchloride/olefin ratio/(mol/mol), [2, 3]; x_3, Reaction temperature/(°C), [0, 25]; x_4, Type of Lewis acid, [$FeCl_3$, $AlCl_3$].

Table 14. (Continued)

Entry	y_1	y_2	y_3	y_4	y_5	y_6	y_7	y_8	y_9	y_{10}	y_{11}	t_1
1	1.0	1.0	0	72.7	9.0	0.3	4.4	0	6.0	0	0	31.79
2	0	0	0	61.3	11.0	0.5	0.2	7.6	0.7	12.0	0	23.23
3	0	0	0	59.4	10.8	0.5	0.5	10.7	0.3	15.3	0	22.32
4	0	0	0	70.5	12.9	0.7	0.1	4.5	0.4	4.9	0	28.58
5	0.9	0.7	0	75.5	8.8	1.3	1.3	0.3	4.1	0.4	0	34.41
6	0	1.0	0	76.3	8.3	1.7	0	5.2	0	4.8	0	35.73
7	0.1	0.7	0	69.4	8.5	1.0	0.5	7.3	0.7	5.5	0	30.50
8	0.5	0.8	0	68.9	8.3	1.0	1.0	0.3	6.6	0.6	0	22.59
9	2.8	2.1	23.1	0	47.2	0	0.1	0.1	0.1	0	0	−51.40
10	1.0	3.1	0	42.4	30.7	5.5	6.1	0.9	0.1	0	0	− 4.25
11	7.5	0	9.8	0	77.3	0	0	0	0	0	0	−67.01
12	0	0.7	0	29.0	7.4	2.8	1.5	23.5	2.5	28.7	0	2.66
13	16.8	0.4	18.5	0	57.0	0.4	0.2	0.1	0.2	0.1	0	−58.09
14	0	0.6	0	33.1	9.7	5.5	2.0	9.7	2.1	14.8	0	2.56
15	19.0	0.3	16.1	0	55.7	0	0	0.8	0	0	0	−57.08
16	0.3	0.8	0	34.7	11.0	6.2	2.1	14.4	1.6	17.7	0	9.54
P_1[c]	0.128	−0.001	−0.187	0.775	−0.581	−0.010	0.008		0.067	0.013	0.075	*	*

[b] Response variables: y_i is the yield (%) of the pyridine "i" in Scheme 8.
[c] PC loadings.

interaction model. A normal probability plot of the estimated parameters is shown in Fig. 10.

It is seen that the variables x_1 (amount of acetyl chloride) and x_4 (type of Lewis acid) as well as their interaction are the only variables which have a significant influence on the product distribution. For other examples of this technique, see [69].

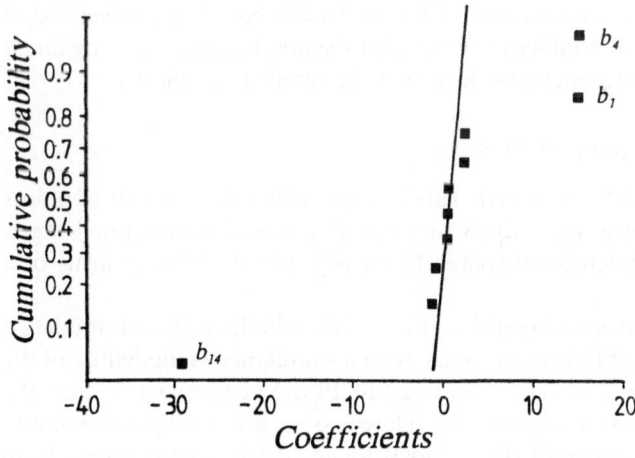

Fig. 10. Normal probability plot of estimated model parameters after fitting a second-order interaction model to the score vector of the pyridines

5.2 PLS: Projections to Latent Structures

PLS is a modelling and computational method for establishing quantitative relations between blocks of variables. Such blocks may, for instance, comprise a block of descriptor variables of a set of test systems (**X** block) and a block of measured responses obtained with these sytems (**Y** block). A quantitative relation between these blocks will make it possible to enter data, **x**, for a new systems and make predictions of the expected responses, **y**, for these systems.

A PLS model can also be established in the reversed sense, i.e. for finding conditions **x** which yield a desired response profile **y**.

The PLS method was developed by H. Wold [71] and has been further developed and extended to chemical systems by S. Wold [72]. A tutorial paper on the method is given in [73]. An important area of application of PLS is in the field of quantitative structure-activity relations [64a, 74]. The applications of PLS for multivariate calibration in analytical chemistry has been thoroughly treated by Martens and Naes [75]. As we shall see, PLS is also a very useful in organic synthesis.

5.2.1 Basic Principles of PLS

PLS is a method which is based on projections, similar to PCA. The underlying principles are easy to understand from a geometrical description of the method.

The blocks of variables are given by the matrices **X** and **Y**. The following notations will be used: The number of objects (e.g. experiments) in both blocks is given by n; the number of variables in the **X** blocks is given by k, $\{x_1 \ldots x_k\}$; the number of variables in the **Y** block is given by m, $\{y_1 \ldots y_m\}$. The **X** block is a $(n \times k)$ matrix, the **Y** block is a $(n \times m)$ matrix.

It is a common misunderstanding that a statistical analysis is only possible when the number of experiments exceeds the number of variables. This is true of multiple regression, but it is note true of PLS. As PLS is based on projections, it can handle any number of variables provided that the number of underlying *latent variables* (cf. principal components) is less than the number of objects.

5.2.2 Geometrical Description of PLS

As was shown in Sect. 4.3.1, any matrix can be represented by a swarm of points in a multidimensional space. Thus, the **X** matrix will define a k-dimensional space, and the **Y** matrix a m-dimensional space. The n objects will define swarms of n points in each space.

In PLS modelling, as much as possible of the variation in the **Y** space is modelled through projections to PLS(Y) components, with a simultaneous modelling of the variations in the **X** space through projections to PLS(X) components, with the constraint that these variations should be related to each other by a maximum correlation between the scores of the **Y** block (denoted u_i) and the scores from the **X** block (deontes t_i) for each PLS dimension i. Fig. 11 illustrates the principles for one PLS dimension (a one-component PLS model).

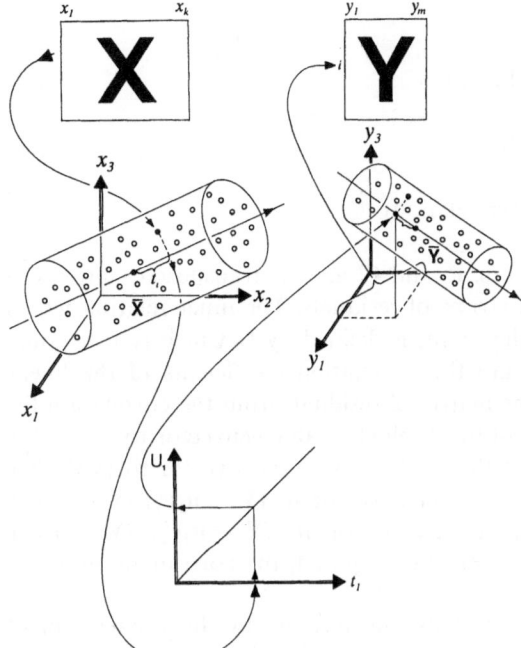

Fig. 11. Geometric illustration of PLS modelling

Predictions for a new object by the PLS model can be understood from Fig. 11 as follows: A new object defines a point, "i" in the **X** space and its projection on the PLS(X) component gives the score t_{1i}. The corresponding score, u_{1i}, along the PLS(Y) component is obtained from the correlation (called the *inner relation*) between t_1 and u_1. The calculated u_{1i} score corresponds to a point along the PLS(Y) component, and the coordinates of this point yield the predicted values $\hat{y}_j$ of each response.

Often, more than one PLS dimension is necessary to account for the systematic variation in the **Y** space. As for PCA, the PLS dimensions can be peeled off, one dimension at a time, until the systematic variation has been described. The models can be established by cross-validation to ensure valid predictions.

The residual variance after fitting the model can be used to compute tolerance limits around the model. This is illustrated by cylinders in Fig. 11. For obtaining reliable predictions with new objects, these should be within the the tolerance limits in the **X** space.

5.2.3 Mathematical Description of PLS

X and **Y** denote the scaled and mean centered matrices of the variations in the **X** and **Y** space, respectively. Scaling to unit variance is usually employed. PLS modelling involves the factorization of the matrices **X** and **Y** into matrices of scores and loadings. In matrix notation, the model

Rolf Carlson and Åke Nordahl

is expressed:

$$X = TP' + E \qquad \text{(X block)} \qquad\qquad (49)$$

$$Y = UC' + F \qquad \text{(Y block)} \qquad\qquad (50)$$

$$U = DT + H \qquad \text{(inner relation)} \qquad\qquad (51)$$

T and U are matrices of score vectors, P' and C' are the transposed matrices of the loading vectors, E and F are matrices of residuals. The inner relation which describes the correlation between the scores is defined by D which is a diagonal matrix in which the elements, d_{ii}, are the correlation coefficients of the linear relation between the scores. H is the matrix of residuals from the correlation fit.

In PLS, the first score vector, t_1, of the X block is an eigenvector to $(XX'YY')$, whereas the first score vector, u_1, of the Y block is an eigenvector to $(YY'XX')$. For the second PLS dimension, t_2 is an eigenvector to $(Y - u_1c_1')$ $(Y - u_1c_1')'$ $(X - t_1p_1')$ $(X - t_1p_1')'$, and u_2 is an eigenvector to $(Y - u_1c_1')$ $(Y - u_1c_1')'$ $(X - t_1p_1')'$. If still more PLS dimensions are required, the corresponding score vectors are defined analogously.

It is, however, not necessary to compute eigenvectors by diagonalization of matrices. The PLS algorithm is based upon the NIPALS algorithm which makes it possible to iteratively determine one PLS dimension at a time. For details of the PLS algorithm, see [75].

Recently, PLS modelling involving non-linear inner relation has been described [76].

5.3 Examples of the Use of PLS Modelling in Synthesis

5.3.1 Prediction of Optimum Conditions for New Substrates in the Willgerodt-Kindler Reaction [19]

The example shows a sequential approach to the study of discrete variations. The Willgerodt-Kindler reaction was studied with a set of *para* substituted aceto-phenones, see Schema 9. As the reaction failed with strong electron-withdrawing substituents, the study was limited to include donor, alkyl, and halogen substituents.

An initial set of five *para* substitutes (H, Me, Cl, MeO, Me$_2$N) acetophenones was selected. The experimental conditions which afforded the maximum yield were determined for each substrate by response surface methods. The optimum settings

54

of the experimental variables, and the observed maximum yield were used as response variables (the **Y** block) in the PLS model, see Table 15.

To characterize the substituents (the **X** block) a set common substituent parameters were compiled from the literature. These parameters were: σ_I (Taft inductive parameter), σ_p (Hammett parameter for *para* substituents), F and R (Swain-Lupton dual substituent parameters), E_S and E_Sc (Taft steric parameters), *van der Waals radius*, L, B_1, B_2, B_3, B_4 (Verloop sterimol parameters), MR (molar refractivity), and π (Hansch lipophilicity parameter). Data are given in Refs. [1, 19] and are not reproduced here.

Table 15. Optimum conditions in the Willgerodt-Kindler reaction of substituted acetophenones

Entry	Subst.	Predicted optimum[a]			Method	Yields (%)	
		S	N	T		y_{Pred}	y_{Obs}^{b}
1	Me	9.6	8.4	133	RSM	96.0	96
2	MeO	9.3	8.9	130	RSM	98.2	100
3	H	7.5	10.3	123	RSM	90.8	94
4	Cl	9.7	9.9	119	RSM	95.0	100
5	Me$_2$N	8.8	8.3	122	RSM	89.0	89
6*	F	7.8	11.0	112	PLS	89.3	–
6	F	8.3	10.6	116	RSM	94.0	93
7*	Br	10.4	9.3	123	PLS	98.4	–
7	Br	10.2	9.5	121	RSM	95.0	95
8	MeS	10.4	8.4	124	PLS	97.0	95

[a] S, The amount of sulfur/ketone (mol/mol); N, the amount of morpholine/ketone (mol/mol); T, the reaction temperature (°C). [b] The experimental yield was determined by gas chromatography (internal standard technique) directly on the reaction mixture.

A first PLS model was determined from Entries 1–5 in Table 15. The model was then used to predict the optimum conditions for *p*-fluorocatophenone, Entry 6*. To check the predictions, the true optimum was determined by response surface modelling, Entry 6. These results were then appended to Entries 1–5, and an upgraded PLS model was determined, which was used to predict the optimum conditions for *p*-bromoacetophenone, Entry 7*. The predictions were quite good as is seen from the validation by response suface modelling, Entry 7. The PLS model was upgraded by including also Entry 7 and used to predict the optimum conditions for *p*-thiomethylacetophenone. Experiments carried out as predicted by the PLS model afforded 95% yield (predicted 97%). This was assumed to be quite close to the true optimum conditions.

5.3.2 Factors Controlling the Regioselectivity in the Fischer Indole Synthesis

The Fischer indole reaction has been extensively used for the synthesis of indole derivatives [77]. A problem is that two isomeric indoles are possible when the reaction is applied to arylhydrazones from dissymmetric methylene ketones, see Scheme 10.

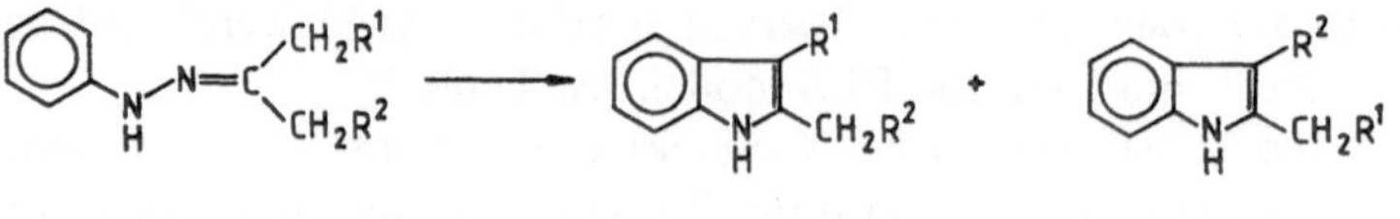

Rolf Carlson and Åke Nordahl

The reaction is acid catalyzed and both Brönsted and Lewis acids have been used. It has been claimed in the literature [78] that the choice of a proper catalyst would be a solution to the selectivity problem. However, the literature is contradictory and bewildering on this point [79].

To clarify whether certain combinations of catalysts and solvents would afford a general selectivity, the study described here was untertaken [80]. Five dissymmetric methylene ketones, twelve Lewis acid catalysts, and ten solvents were selected to afford a fairly uniform spread in the principal properties score plots, see Fig. 12.

The selected items give a total of 600 possible combinations. Of these, 296 systems were studied in experimental runs, and of these, 162 afforded the Fischer indole reaction. The other systems failed. The successful systems were used for PLS modelling. The reaction were monitored by gas chromatography for 48 h after which time the increase in yield was insignificant. No isomerization occurred. The isomer distribution was determined from the gas chromatograms. The response used for PLS modelling was the regioisometrc excess, RE = Amount of major isomer − Amount of minor isomer.

The reaction systems were characterized by the corresponding principal propery scores. For the ketones, the Charton v parameter [81] for the substituents onf the α and α' carbons were also included to take possible steric effects into account. The descriptors are summarized in Table 16.

To define the **X** block, the cross-products and the squares of the descriptors were also included to take possible interaction effects and non-linear effects into account. This gave a total of 44 variables in the **X** block.

PLS modelling was accomplished in two steps: In a first step, a PLS model was determined from 124 reaction systems. This model was then used to predict the regioselectivity of the remaining 38 systems. A correlation coefficient of 0.93 was obtained for the correlation between $y_{Predicted}$ and $y_{Observed}$. This was considered to be satisfactory result. The final PLS model was determined from all 162 systems. A three-components model was significant according to cross-validation and accounted for 86.4% (64.1 + 16.3 + 6.4%) of the total variance of the response.

To determine which properties of the reaction systems contribute to the observed selectivity, plots of the PLS weights of the x variables were used in combination with a statistical criterion, *modelling power* [43]. We will omit these details, full accounts are given in Refs. [1, 80].

The most important factors for controlling the selectivity were, as expected, the properties of the ketone substrate. The steric parameters were the most important,

Fig. 12. Score plots used for selecting test items in the Fischer indol synthesis. Selected ▶ items: **(a)** ketones, **(b)** Lewis acids, **(c)** solvents are marked with *dots*. For identification numbers in the score plots, see Refs. [51, 55, 61]

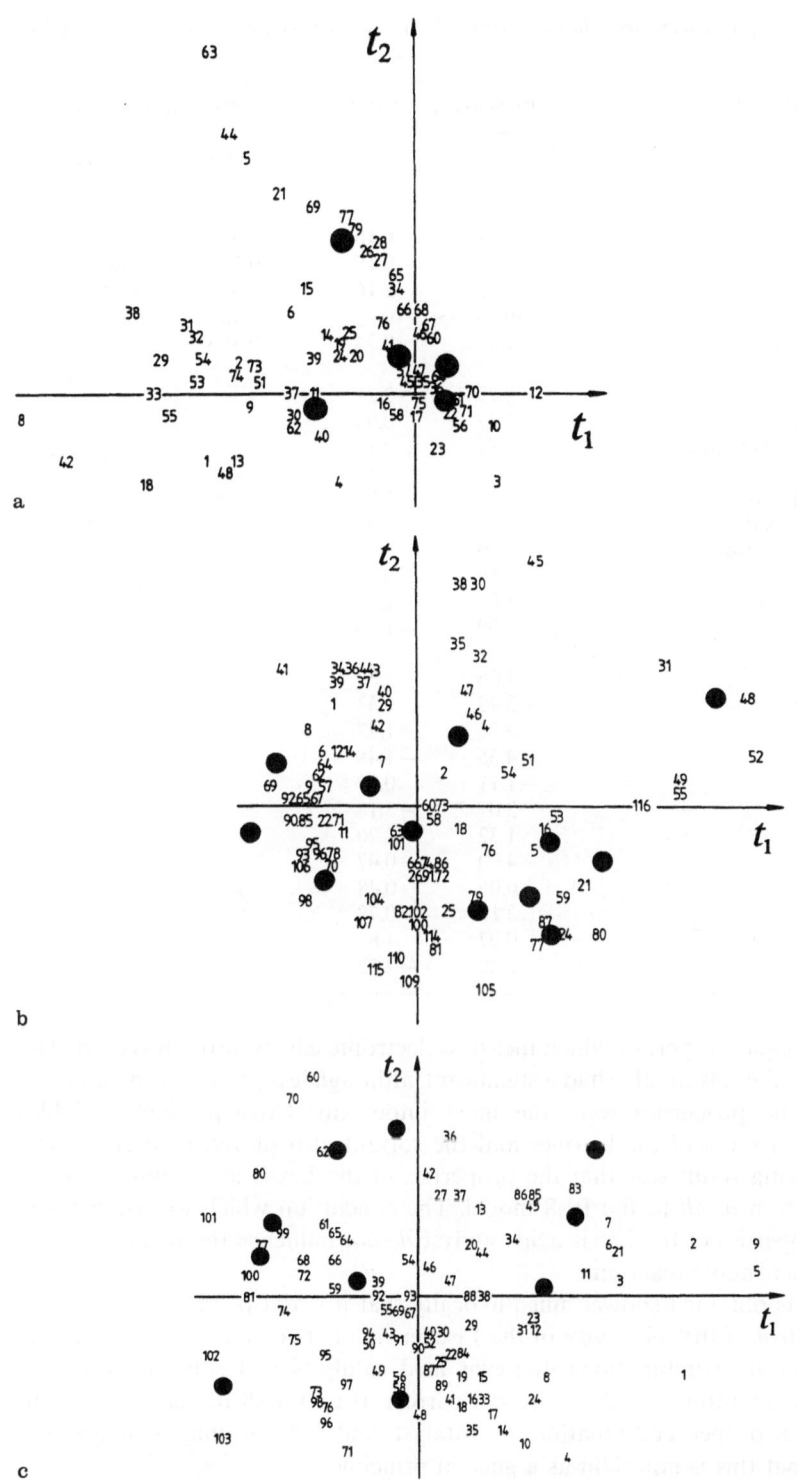

Table 16. Descriptors used for characterizing the reaction systems in the Fischer indole reaction

System constituent	Principal properties		Steric parameter, v	
	t_1	t_2	R^1CH_2	R^2CH_2
Ketones:				
3-Hexanone	2.34	−0.16	0.68	0.56
2-Hexanone	2.46	−0.15	0.52	0.68
3-Undecanone	−0.52	2.16	0.68	0.56
1-Phenyl-2-butanone	−0.74	−0.42	0.56	0.70
5-Methyl-3-heptanone	1.17	0.74	0.56	1.00
Solvents:				
Sulfolane	3.29	2.21		
Carbon disulfide	−3.27	0.98		
N,N-Dimethylacetamide	1.81	0.15		
Quinoline	−0.29	3.13		
1,2-Dichlorobenzene	−0.99	2.27		
Dimethyl sulfoxide	2.54	0.95		
Carbon tetrachloride	−2.49	0.38		
Chloroform	−1.56	−0.59		
Tetrahydrofuran	−1.04	−1.57		
Hexane	−3.00	−1.20		
Lewis acids:				
BF_3	7.05	2.80		
CuCl	−3.42	1.33		
ZnI_2	−2.25	−1.77		
$TiCl_4$	4.35	−1.48		
$ZnCl_2$	−1.11	0.79		
$SbCl_5$	3.07	−3.18		
PCl_3	1.32	−2.70		
CuI	−4.00	−0.47		
$FeCl_3$	−0.06	−0.48		
$SiCl_4$	3.18	−0.78		
$AlCl_3$	0.97	1.83		
$SnCl_4$	2.20	−2.22		

but the principal properties which measure electronic effects also intervened. The properties of the solvent also had a significant, although less pronounced influence. The lipophilic properties were the most important. Cross-product variables between descriptors of the ketones and the solvent also played their roles. The most surprising result was that the properties of the Lewis acid catalysts made no contribution *at all* to the PLS model. The conclusion which was drawn was that the properties of the Lewis acid catalyst *do not* influence the regioselectivity of the Fischer indole reaction.

The results obtained showed unequivocally that it is not possible to achieve a general control of the selectivity of the Fischer indole reaction *in solution* by the choice of certain combinations of Lewis acid catalysts and solvents. Previous claims in the literature are therefore not correct. It may well be that, for certain substrates, a proper combination of catalyst and solvent may enhance the selectivity, but this is not valid as a general principle.

The results pointed in another direction. The effects of steric congestion in the substrates are the most important factor. One way to enhance the selectivity would therefore be to amplify the steric effects by forcing the reaction to occur close to or on the surface of a solid acid catalyst. This was found to be a valid conclusion, and by using zeolites as catalysts the mode of ring closure could be partially controlled by a proper choice of zeolite. For certain substrates, the regioselectivity was reversed by changing the catalyst from Mordenite to Zeolite-Y [82].

6 Conclusions

When a chemical reaction is elaborated into a synthetic method, an extensive amount of work is usually necessary to clarify the questions which may arise. Any scientific experiment is carried out in a framework of known facts and mere assumptions. A danger when the results of an experiment are interpreted is that assumptions which later on turn out to be false might interfere with the conclusions drawn. This problem is imminent when a newly discovered reaction or when *ideas* for new reactions are subjected to experimental studies. If mere speculations on the reaction mechanisms are the sole basis for the selection of test systems or for determining the experimental conditions, the choice may be too narrow, and there is a risk that a useful new reaction is overlooked.

The methods and examples outlined in this chapter show how multivariate statistical tools may assist in this context, both for the exploration of the reaction space and in the experimental space. Proper designs ensure that sufficiently large variations of the experimental systems are covered by the experiments. The number of necessary experiments can be kept low by a proper choice of experimental design. A properly designed experiment permits systematic variations of the experimental results to be modelled so that these variations can be traced to the perturbations of the experimental system.

We know that organic synthesis by some practioners in the field is seen as an "artistic" discipline. It is sometimes even claimed that statistical methods have no roles to play in this area, since chemical knowledge and chemical intuition would guide the chemist to do the right things. We think this is a wrong attitude and, hopefully, based upon a misunderstanding of what statistics can and should do in chemistry. It is true, that knowledgeable chemists do excellent chemistry. It is, however, also true that in the literature there is abundance of mediocre chemistry. The multivariate methods outlined in this chapter can be seen as methods for quantifying chemical intuition. However, statistics can never substitute chemical knowledge. Chemical theory should be used in the context where it is powerful. The includes all aspects when a problem is formulated so that its relations to known facts are fully realized. It is also important for an initial experimental set-up and for determining how the results should be analyzed. Chemical theory should, however, never be used for the immediate evaluation of the observed experimental result. Owing to an omnipresent random experimental error, all experimental data *must* be evaluated by statistical means so that real and systematic variation of the experimental results can be distinguished from mere random variations. Statistically significant variations must then be scrutinized in the light of chemical theory to determine whether they are also chemically significant. Chemical theory and statistics are therefore complementary and should always be used in combination whenever a chemical *experiment* is designed to solve a chemical problem.

The strategies for design and analysis of organic synthetic experiments discussed in this chapter have been tailored to find answers to the following questions: *Which* factors have influence on the observed result? *How* do these factors exert their influence? Answers to these questions will often be sufficient for establishing an optimum synthetic method. The answers will form a good starting point for designing experiments with a view to answering also the *Why?* questions. Any proposals on the mechanistic level must explain why certain factors are important while others are not.

Acknowledgements. Financial support from the Swedish Natural Science Research Council and from the Swedish National Board for Industrial and Technical Development is gratefully acknowledged. We thank Elsevier Science Publishers for the kind permission to reproduce Figs. 1, 8 and 11 from Ref. [1]. The authors wish to express their gratitude to past and present coworkers who have made significant contributions to topics treated, and whose names appear in the reference list. The authors also thank Doc. Michael Sharp for linguistic assistance.

7 Appendix: Computer Programs

In this appendix are listed some commercially available computer programs which can be used for the methods discussed above. The list is by no means complete. We have only included such programs which we know are useful, either from our own experience, or from the experience of trustworthy colleagues in the field.

For multiple linear regression, there are several hundreds of programs available on the market, but in the list below we have included only such programs which can assist in response surface modelling and which have routines for plotting the fitted surfaces.

For PCA and PLS modelling, we have included only such programs which employ the NIPALS algorithm. The reason for this is that such programs can handle situations where there are missing data, an almost inevitable problem when data are recorded in real experiments.

Programs for response surface modelling
For personal computers (IBM PC XT/AT and compatibles):

MODDE
Umetri AB,
P.O. Box 1456, S-901 24 Umeå, Sweden

NEMROD
LPRAI
Att. Prof. R. Phan-Tan-Luu
Université d'Aix-Marseille III
Av. Escadrille Normandie Niemen
F-13397 Marseille Cedex 13, France

STATGRAPHICS
STSC, Inc.
2115 East Jefferson Street
Rockville, MD 20852, USA

ECHIP
Expert in a Chip, Inc.
RD! Box 384Q
Hockesin, DE 19707, USA

XSTAT
Wiley Professional Software
John Wiley and Sons, Inc.
605 Third Avenue
New York, NY 10158, USA

For mini and mainframe computers:
RS1/Discover
BBN Software Products Corp.
10 Fawcett Street
Cambridge, MA 02238, USA

SAS
SAS Institute, Inc.
SAS Circle
P.O. Box 800
Cary, NC 27512, USA

Programs for PCA and PLS modelling
For personal computers (IBM PC XT/AT and compatibles):
SIMCA-R (V. 4.3)
Umetri AB
P.O. Box 1456
S-90124 Umeå, Sweden
MDS, Inc.
371 Highland Ave.
Winchester, MA 01890, USA

SIRIUS
Chemical Institute
Att. O.M. Kvalheim
University of Bergen
N-3007 Bergen, Norway

Rolf Carlson and Åke Nordahl

UNSCRAMBLER II
Camo AS
Jarleveien 4
N-7041 Trondheim, Norway

PIROUETTE
Info Metrix
220 Sixth Avenue, Suite 8333
Seattle, WA 98121, USA.

8 References

1. Carlson R (1992) Design and optimization in organic synthesis. Elsevier, Amsterdam
2. For examples, see (a) Corey EJ (1971) Quart. Rev. 25: 455; (b) Wipke WT, Howe WJ (eds) (1987) Computer-assisted organic synthesis (ACS Symposium Series 61); (c) Barone R, Chanon M (1986) In: Vernin G, Chanon M (eds) Computer aids to chemistry. Ellis Horwood, Chichester
3. For examples of the Ugi approach, see (a) Gasteiger J, Jochum C (1978) Topics Curr Chem 74: 94; (b) Bauer H, Herges R, Fontain E, Ugi I (1985) Chimia 39: 43; (c) Ugi I, Wochner M (1988) J Molec Struct (Theochem) 165: 229
4. For examples, see Zefirov NS (1987) Acc Chem Res 20: 237
5. Box GEP, Hunter WG, Hunter JS (1978) Statistic for experimenters, Wiley, New York
6. For examples, see Laird ER, Jorgensen WL (1990) J Org Chem 55:9 and earlier papers in this series
7. Draper NR, Smith H (1981) Applied regression analysis, 2nd edn. Wiley, New York
8. Box GEP, Wilson KB (1951) J Roy Statist Soc B 13: 1
9. For details, see (a) Spendley W, Hext GR, Himsworth FR (1962) Technometrics 4: 441; (b) Nelder JA, Mead R (1964) Computer J 7: 308; (c) Burton KWC, Nickless G (1987) Chemometr Intell Lab Syst 1: 135; (d) Morgan E, Burton KW, Nickless G (1990) Chemometr Intell Lab Syst 7: 209; (e) Morgan E, Burton KW, Nickless G (1990) Chemometr Intell Lab Syst 8: 97
10. Cochran WG, Cox GM (1957) Experimental designs. Wiley, New York
11. (a) Baynes CK, Rubin IB (1986) Practical experimental design and optimization methods in chemistry. VCH Publishers, Derrisfield Beach; (b) Diamond WJ (1981) Practical experimental designs for engineers and scientists. Van Nostrand Reinold, New York; (c) Haaland PD (1989) Experimental design in biotechnology. Marcel Dekker, New York; (d) Biles WE, Swain JJ (1980) Optimization and industrial experimentation. Wiley, New York; (e) Deming SM, Morgan SL (1987) Experimental design. A chemometric approach. Elsevier, Amsterdam; (f) Morgan E (1991) Chemomerics: Experimental design. Wiley, Chichester
12. Plackett RL, Burman JP (1946) Biometrika 33: 305, 328
13. (a) Mitchell TJ (1974) Technometrics 16: 203; (b) Cook RD, Nachtsheim CC (1980) Technometrics 22: 315
14. Fedorov VV (1972) Theory of optimal experiments, Academic, New York
15. Daniel C (1959) Technometrics 1: 311; (b) Daniel C (1976) Application of statistics to industrial experimentation. Wiley, New York
16. Nilsson Å (1984) Syntes och reduktion av enaminer, Thesis, Umeå University, Umeå
17. (a) Carlson R, Nilsson Å (1985) Acta Chem Scand B 39: 181; (b) Nilsson Å, Carlson R (1985) Acta Chem Scand B 39: 187
18. Godawa C (1984) Etude de l'hydrogenation catalytique du furanne: Optimisation du procede, Thesis, L'Institut National Polytechnique de Toulouse, Toulouse
19. Carlson R, Lundstedt T, Shabana R (1986) Acta Chem Scand B 40: 534
20. Carlson R, Lundstedt T, Shabana R (1986) Acta Chem Scand B 40: 694
21. Lundstedt T, Carlson R, Shabana R (1986) Acta Chem Scand B 41: 157

22. Carlson R, Lundstedt T (1987) Acta Chem Scand B 41: 164
23. Carlson R, Phan-Tan-Luu R, Mathieu D, Ahouande FS, Babadjamian A, Metzger J (1978) Acta Chem Scand B 32: 335
24. (a) Taguchi K, Westheimer FH (1971) J Org Chem 36: 1570; (b) Roelofsen DP, van Bekkum H (1972) Recl Trav Chim Pays-Bas 91: 605
25. (a) Box GEP, Wilson KB (1951) J Roy Statist Soc B 13: 1; (b) Box GEP (1954) Biometrics 10: 16; (c) Box GEP, Youle PV (1955) Biometrics 11: 287
26. (a) Myers RM (1971) Response surface methodology. Allyn and Bacon, Boston; (b) Box GEP, Draper NR (1987) Empirical model-building and response surfaces. Wiley, New York
27. (a) Box GEP, Behnken DW (1969) Technometrics 2: 455; (b) Hoke TA (1974) Technometrics 17: 375
28. Doehlert DH (1970) Appl Statist 19: 231
29. Box GEP, Hunter JS (1957) Ann Math Statist 28: 195
30. Carlson R, Nilsson Å, Strömqvist M (1983) Acta Chem Scand B 37: 7
31. White WA, Weingarten H (1967) J Org Chem 32: 231
32. Carlson R, Nilsson Å (1984) Acta Chem Scand B 38: 49
33. Hansson L, Carlson R (1989) Acta Chem Scand 43: 1888
34. See Ref [1], pp 284–290
35. Hansson L, Carlson R (1989) Acta Chem Scand 43: 304
36. Pearson K (1901) Phil Mag 2: 559
37. Hotelling H (1933) J Educ Psych 24: 417, 498
38. Thurnstone LL (1947) Multiple factor analysis. Chicago University Press, Chicago
39. Golub G, Van Loan C (1983) Matrix computations. The Johns Hopkins University Press, Oxford
40. (a) Karhunen K (1947) Ann Acad Sci Fennicae Ser A: 137; (b) Loeve M (1948) Processes stochastiques et mouvements brownien. Herman, Paris
41. Simonds JL (1963) J Opt Sci Am 53: 968
42. Malinowksi ER, Howery DC (1980) Factor analysis in chemistry. Wiley, New York
43. Wold S, Sjöström M (1977) In: Kowalski BR (ed) Chemometrics: Theory and application (ACS Symposium Series 52) 243
44. (a) Gampp H, Maeder M, Meyer Cl, Zuberbühler AD (1985) Talanta 32: 1133; (b) for a tutorial, see Keller HR, Massart DL (1992) Chemometr Intell Lab Syst 12: 209
45. Kvalheim OM; Liang Y-z (1992) Anal Chem 64: 936
46. Weiner P (1972) Can J Chem 50: 448
47. (a) Jolliffe LT (1986) Principal component analysis. Springer, Berlin Heidelberg New York; (b) Wold S, Esbensen K, Geladi, P (1987) Chemometr Intell Lab Syst 2: 37
48. Zalewksi RI (1990) In: Taft RW (ed) Progr Phys Org Chem 18: 77
49. Wold H (1966) Research papers in statistics: Festschrift for J Neyman, Wiley, New York, p 411
50. Wold S (1978) Technometrics 20: 397
51. Carlson R, Lundstedt T, Nordahl Å, Prochazka M (1986) Acta Chem Scand B 40: 522
52. Reetz M, Maier WF, Heimbach H, Giannis A, Anastassious G (1980) Chem Ber 113: 3734
53. McCulloch AW, Smith DG, McInnes AG (1973) Can J Chem 51: 4125
54. Martin HL, Fieser LF (1943) Org Synth Coll Vol 2: 569
55. Carlson R, Lundstedt T, Albano C (1985) Acta Chem Scand B 39: 79
56. See Ref. [1], pp 390–395
57. (a) Bohle M, Kollecher W, Martin D (1977) Z Chem 17: 161; (b) Elguero J, Fruchier A (1983) Ann Quim Ser C 79: 72; (d) Svoboda P, Rytella O, Vecera M (1983) Coll Czech Chem Commun 48: 3287; (d) Maria PC, Gal J-F, de Francheschi J, Fargin E (1987) J Am Chem Soc 109: 483; (e) Cramer III RD (1980) J Am Chem Soc 102: 1837, 1849; (f) Chastrette M (1979) Tetrahedron 35: 144; (g) Chastrette M, Carretto J (1982) Tetrahedron 38: 1615; (h) Chastrette M, Rajzamann M, Chanon M, Purcell KF (1985) J Am Chem Soc 107: 1; (i) Zalewski RI, Kokocinska H, Reichardt C (1989) J Phys Org Chem 2: 232

58. Reichardt C (1988) Solvent and solvent effects in organic chemistry. Second, completely revised and enlarged edition, Verlag Chemie, Weinheim
59. Nordahl Å, Carlson R (1988) Acta Chem Scand B 42: 28
60. Prochazka MP, Carlson R (1989) Acta Chem Scand 43: 651
61. Carlson R, Prochazka MP, Lundstedt T (1988) Acta Chem Scand B 42: 145
62. Patai S (ed) (1966) The Chemistry of the Carbonyl Group, Wiley, London
63. Carlson R, Prochazka MP, Lundstedt T (1988) Acta Chem Scand B 42: 157
64. For examples, see (a) Hellberg S, Sjöström M, Skagerberg B, Wold S (1987) J Med Chem 30: 1126; (b) Hellberg S, Sjöström M, Would S (1986) Acta Chem Scand B 40: 135
65. For examples, see (a) Jonsson J, Eriksson L, Sjöström M, Wold S, Tosato ML (1989) Chemometr Intell Lab Syst 5: 169; (b) Eriksson L, Jonsson J, Sjöström M, Wold S (1989) Chemomtr Intell Lab Syst 7: 131
66. Kaufmann P (1990) Applications of chemometrics in the analysis and characterization of lipids, Thesis, Stockholm University, Stockholm
67. Alunni S, Clementi S, Edlund U, Johnels D, Hellberg S, Wold S (1983) Acta Chem Scand B 37: 47
68. Box GEP, Hunter WG, McGregor JF, Erjavec R (1973) Technometrics 15: 33
69. Carlson R, Nordahl Å, Barth T, Myklebust R (1992) Chemometr Intell Lab Syst 12: 237
70. Diallo A, Hirschmueller A, Arnaud M, Roussel C, Mathieu D, Phan-Tan-Luu R (1983) Nouv J Chim 7: 433
71. Wold H (1982) in Jöreskog K-G, Wold H (eds) Systems under indirect observations, vols I, II. North Holland, Amsterdam
72. For examples, see (a) Wold S, Ruhe A, Wold H, Dunn III WJ (1984) SIAM J Scient Statist Comput 5: 735; (b) Wold S, Albano C, Dunn III WJ, Edlund U, Esbensen K, Geladi P, Hellberg S, Johansson E, Linberg W, Sjöström M (1984) In: Kowalski 111 (ed) Chemometrics: Mathematics and statistics in chemistry. Riedel, Dordrecht; (c) Wold S, Albano C, Dunn III WJ, Esbensen K, Geladi P, Hellberg S, Johansson E, Linberg W, Sjöström M, Skagerberg B, Wikström C, Öhman J (1989) In: Brandt J, Ugi I (eds) Computer applications in chemical research and education, Hühtig Verlag, Heidelberg
73. Geladi P, Kowalski BR (1986) Anal Chim Acta 185: 1
74. For examples, see (a) Hellberg S, Sjöström M, Skagerberg B, Wikström C, Wold S (1987) In: Hadži D, Jerman-Blažič B (eds) QSAR in drug design and toxicology. Elsevier, Amsterdam, p 255; (b) Wold S, Berntsson P, Eriksson L, Geladi P, Hellberg S, Johansson E, Jonsson J, Kettaneh-Wold N, Lindgren F, Rännar S, Sandberg M, Sjöström M (1991) In: Silipo C, Vittoria A (eds) QSAR: Rational approaches to the design of bioactive compounds, Elsevier, Amsterdam p 15
75. Martens H, Naes T (1989) Multivariate calibration. Wiley, Chichester
76. (a) Wold S, Kettaneh-Wold N, Skagerberg B (1989) Chemometr Intell Lab Syst 7: 53; (b) Frank IE (1990) Chemometr Intell Lab Syst 8: 109; (c) Wold S (1992) Chemometr Intell Lab Syste 14: 71
77. Robinson B (1982) The Fischer indole synthesis. Wiley, Chichester
78. Baccolini G, Bartoli G, Marotta E, Todesco PE (1983) J Chem Soc Perkin Trans 1: 2695
79. See Ref [77], pp 172–393 and references cited therein
80. Prochazka MP, Carlson R (1989) Act Chem Scand 43: 651
81. Charton M (1983) Topics Curr Chem 114: 57
82. (a) Prochazka MP, Eklund L, Carlson R (1990) Act Chem Scand 44: 610; (b) Prochazka MP, Carlson R (1990) Act Chem Scand 44: 614

Enumeration of Benzenoid Chemical Isomers with a Study of Constant-Isomer Series

Sven J. Cyvin, Björg N. Cyvin, and Jon Brunvoll

Division of Physical Chemistry, The University of Trondheim, N-7034 Trondheim-NTH, Norway

Table of Contents

S. J. Cyvin, B. N. Cyvin, and J. Brunvoll

Benzenoid (chemical) isomers are, in a strict sense, the benzenoid systems compatible with a formula $C_nH_s \equiv (n; s)$. The cardinality of C_nH_s, viz. $|C_nH_s| \equiv |n, s|$ is the number of isomers pertaining to the particular formula. The generation of benzenoid isomers (*aufbau*) is treated and some fundamental principles are formulated in this connection. Several propositions are proved for special classes of benzenoids defined in relation to the place of their formulas in the Dias periodic table (for benzenoid hydrocarbons). Constant-isomer series for benzenoids are treated in particular. They are represented by certain C_nH_s formulas for which $|n; s| = |n_1; s_1| = |n_2; s_2| = ...$, where $(n_k; s_k)$ pertains to the k times circumscribed C_nH_s isomers. General formulations for the constant-isomer series are reported in two schemes referred to as the Harary-Harborth picture and the Balaban picture. It is demonstrated how the cardinality $|n; s|$ for a constant-isomer series can be split into two parts, and explicit mathematical formulas are given for one of these parts. Computational results are reported for many benzenoid isomers, especially for the constant-isomer series, both collected from literature and original supplements. Most of the new results account for the classifications according to the symmetry groups of the benzenoids and their Δ values (color excess).

1 Introduction

One of the classical problems in the enumeration of chemical isomers was posed more than a hundred years ago [1, 2]: — How many alkanes with the formula $C_N H_{2N+2}$ can be constructed? It is a well-known fact in organic chemistry that there is one methane ($N = 1$), one ethane ($N = 2$), one propane ($N = 3$), but two butanes ($N = 4$) and three pentanes ($N = 5$). These isomers can be symbolized in the simplest way as:

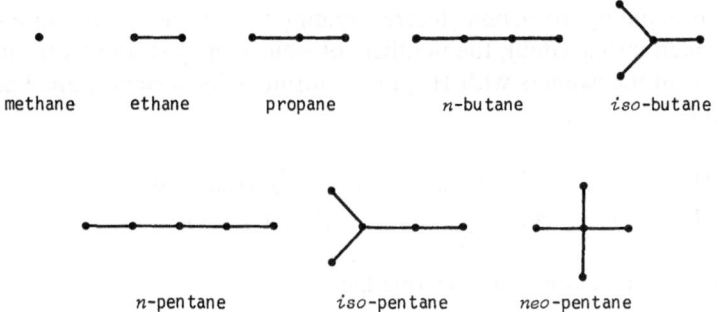

methane   ethane   propane   *n*-butane   *iso*-butane

n-pentane   *iso*-pentane   *neo*-pentane

Even in this old problem the "topological" isomers were counted, viz. the structural formulas, but not the structures themselves; different kinds of structural isomerism were neglected. In modern language this is an enumeration of nonisomorphic chemical graphs [3]. The particular problem on the numbers of alkane isomers in general was solved satisfactorily, more than fifty years after it had been posed, in terms of a recursive algorithm [4]. The numbers represent a typical problem in computational chemistry and has also been attacked as such in more recent times [5–7]. The presumably largest exact number of alkane $C_N H_{2N+2}$ isomers, which has been achieved by computer aid, pertains to $C_{80}H_{162}$ and reads [7]:
10 564 476 906 946 675 106 953 415 600 016
The enumeration of isomers of benzenoid hydrocarbons on a large scale is a relatively new area. Of course it has been known for a long time that there is one benzene C_6H_6 ($h = 1$), one naphthalene $C_{10}H_8$ ($h = 2$), but two benzenoid hydrocarbons $C_{14}H_{10}$ ($h = 3$), viz. anthracene and phenanthrene. Here h is the number of six-membered (benzenoid) rings or hexagons. For $C_{18}H_{12}$ ($h = 4$) there are known to be five isomers:

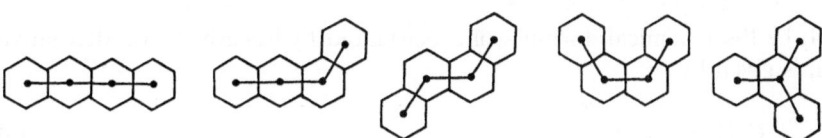

This fact is expressed as

$$|C_{18}H_{12}| = 5 \tag{1}$$

All the above benzenoid hydrocarbons are catacondensed. For this class, corresponding to benzenoids without internal vertices, and represented by the general formula $C_{4h+2}H_{2h+4}$, the numbers of isomers are known up to $h = 15$ (when helicenes are excluded) [8]:

$$|C_{62}H_{34}| = 4799205 \qquad (2)$$

The numbers $|C_{4h+2}H_{2h+4}|$ for $h > 1$ increase steadily with the h values, for $5 \leq h < 15$ by a factor between three and four. On the other hand, if the pericondensed benzenoid hydrocarbons (corresponding to benzenoids with internal vertices) are taken into account, the numbers of isomers may seem to vary in a chaotic way. For all the isomers with H_{14} in the formula, for instance, one has the set of numbers:

$$\begin{aligned} |C_{22}H_{14}| = 12, \qquad |C_{24}H_{14}| = 14, \qquad |C_{26}H_{14}| = 10 \\ |C_{28}H_{14}| = 9, \qquad |C_{30}H_{14}| = 4, \qquad |C_{32}H_{14}| = 1 \end{aligned} \qquad (3)$$

To take a still more picturesque example, one has

$$|C_{600}H_{60}| = 1 \qquad (4)$$

In the following we shall give a full account of the one-isomer series of benzenoids (defined below), which explain the unities in Eq. (4) and the last equation of (3).

The enumeration of benzenoid isomers is the subject of another chapter of this monograph series [8]. It contains a comprehensive historical survey and an extensive tabulation of the enumeration results. In the present chapter some advances of the same topic are reported, but to avoid too much overlapping frequent references to the previous chapter [8] are made. Emphasis will presently be laid on the constant-isomer series of benzenoids (defined below), which is a generalization of the one-isomer series. This topic was only mentioned in the concluding remarks of the previous chapter.

2 Basic Definitions

2.1 Benzenoid Isomers

Let C_nH_s be the (chemical) formula of a benzenoid hydrocarbon. An alternative notation is given by

$$C_nH_s = (n; s) \qquad (5)$$

The cardinality of C_nH_s, written $|C_nH_s|$ or in the alternative notation

$$|C_nH_s| = |n; s| \qquad (6)$$

is the number of benzenoid isomers for the given formula. It is defined as the number of nonisomorphic benzenoid systems compatible with the formula C_nH_s.

Here a benzenoid system (or shortly benzenoid) [9–13] is defined as a geometrically planar, simply connected polyhex. Consequently, the helicenic and coronoid systems are excluded. The coefficients of the formula C_nH_s, viz. n and s, correspond to the total number of vertices and the number of vertices of degree two, respectively, in the benzenoid.

In addition to the more or less general references [9–13] cited above we wish to mention a significant and relatively early paper by Polansky and Rouvray [14]. It contains an explicit treatment of the benzenoid chemical formulas (C_nH_s) and their invariants (n, s). This treatment seems by and large to have been overlooked in later works.

2.2 Addition Modes

Any benzenoid with $h + 1$ hexagons can obviously be generated (or built up) by adding one hexagon to the perimeter of a benzenoid with h hexagons. This principle is very useful in computerized generations of benzenoids. Five types of the additions are distinguished and referred to as [12, 13, 15] (i) one-, (ii) two-, (iii) three-, (iv) four- and (v) five-contact additions. These five additions can be described in the following picturesque way, having in mind the designations of different formations on the perimeter [11, 12]:

(i)   fusion (or annelation) to a free edge;
(ii) filling a fissure;
(iii) embedding in a bay;
(iv) covering a cove;
(v)   immersing in a fjord (fiord).

Also the general definition of hexagon modes [11, 12] is very useful. For our purposes five of these modes, referred to as *addition modes*, are especially important. They are the modes which a hexagon may acquire after it has been added to the perimeter of a benzenoid. For the types (i)–(v) these modes are L_1, P_2, L_3, P_4 and L_5, respectively.

The above features are illustrated in the following diagram.

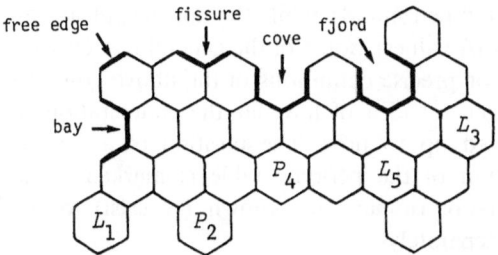

69

In the subsequent treatment three of the addition modes, viz. (i), (ii) and (iii) are especially important. For the sake of convenience they are illustrated below. The added hexagons are grey.

one-contact addition

two-contact addition

three-contact addition

3 Periodic Table for Benzenoid Hydrocarbons

The periodic table for benzenoid hydrocarbons [6, 16] is described elsewhere [8] with numerous relevant references included. It is an array of C_nH_s formulas for benzenoids in the coordinate system (d_s, n_i), where d_s is the Dias parameter [8, 16], and n_i is the number of internal vertices for the benzenoids with the particular formula. The reader is especially referred to the main citation [8] for an explicit listing of the formulas in a considerable part of the table (in the version which also will be used presently) and should also observe the description of the *staircase-like boundary* therein. In Fig. 1 quite a large portion of the periodic table for benzenoid hydrocarbons is displayed as a dot diagram, where each dot represents a formula C_nH_s. All the formulas can actually be retrieved quite easily by means of a recursive algorithm [8] and the initial condition ($C_{10}H_8$ for naphthalene at the upper-left corner). Note that the benzenoids with the same number of hexagons (h) have formulas along a diagonal in the table.

4 Comprehensive Survey of Computational Results

A great deal of work has been done on the enumeration of benzenoid isomers with given formulas C_nH_s or, in other words, the computation of the cardinalities $|C_nH_s|$. Most of these computational results are listed in extensive tables of the review chapter [8] frequently cited above. Those tables contain detailed documentations to the original sources of the data. In all of the cases the cardinalities are separated according to a classification of the benzenoids into Kekuléan and non-Kekuléan systems. Furthermore, in most of the cases, a finer classification is known: the Kekuléans are separated into normal and essentially disconnected benzenoids, while the non-Kekuléans are classified according to the color excess (Δ values). Some of the general references or reviews [8, 11–13] may be consulted for precise definitions of the above concepts.

In Fig. 2 is an attempt to give a general idea of how far the enumerations of benzenoid isomers have been advanced up to now. The position of a formula C_nH_s in the dot-diagram representation of the periodic table is marked by an asterisk if the corresponding numbers of isomers are known (at least) for the Kekuléan and non-Kekuléan systems separately.

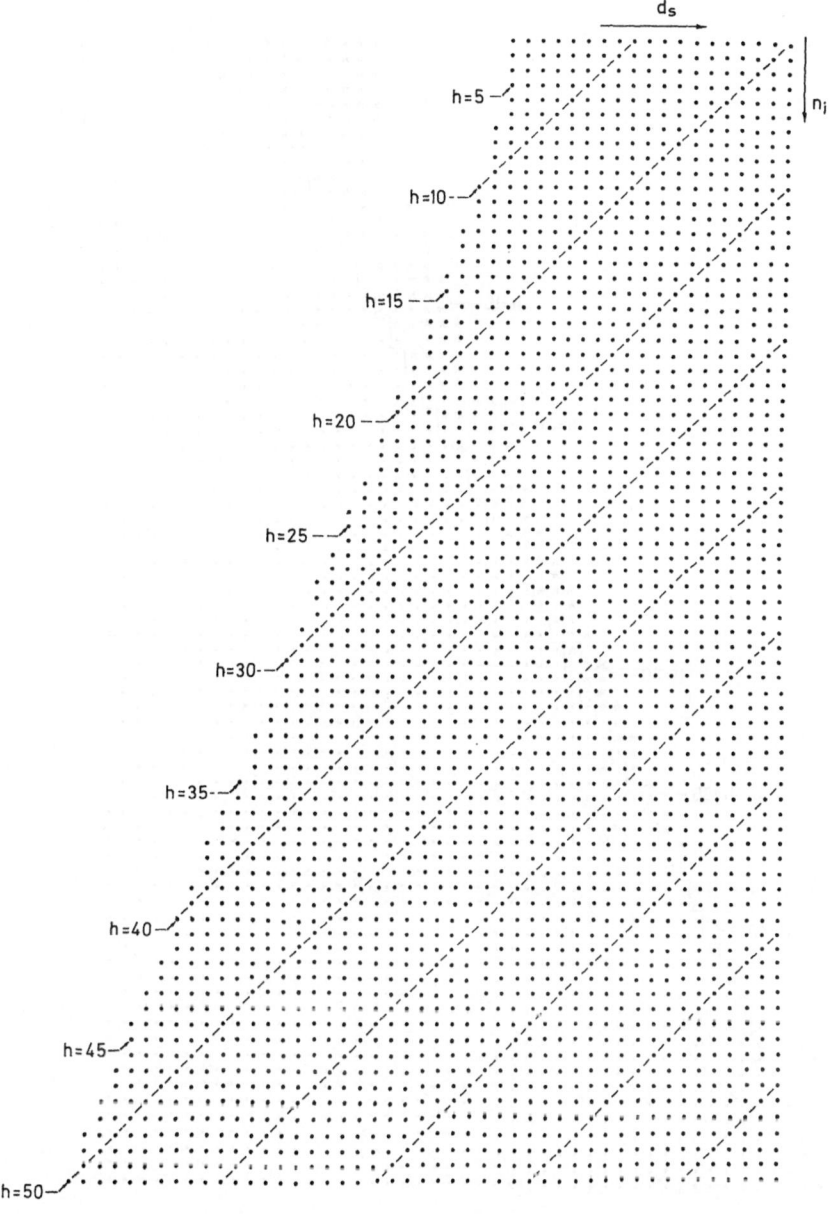

Fig. 1. The periodic table for benzenoid hydrocarbons as a dot diagram

The full-drawn staircase line in Fig. 2 indicates the limit for the significant enumerations of Stojmenović et al. [17]. In that work the separate numbers for Kekuléan and non-Kekuléan systems were not determined. However, it was taken into account, when the asterisks were distributed, that *odd-carbon atom* formulas ($C_n H_s$ with n and s odd) pertain to non-Kekuléans only, while *even-carbon atom*

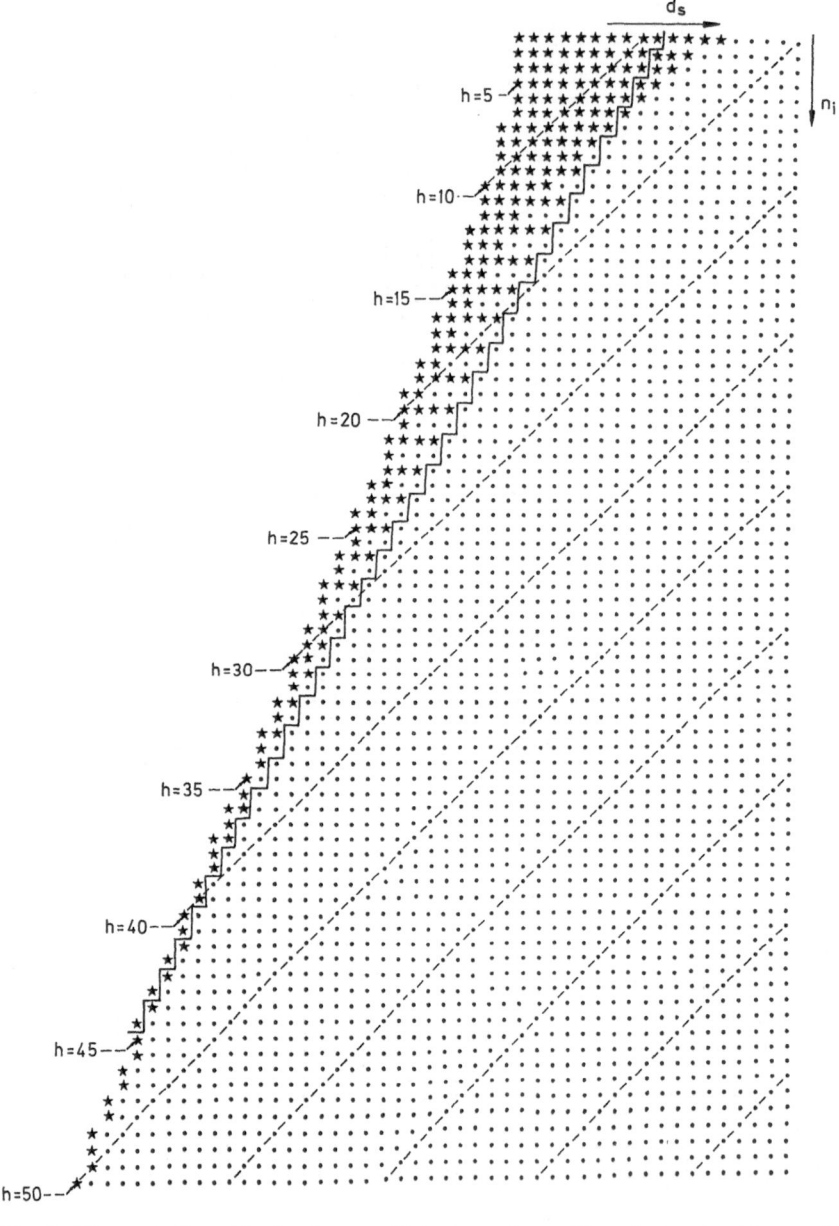

Fig. 2. The periodic table for benzenoid hydrocarbons (dot diagram) with asterisks; an *asterisk* indicates that the numbers of Kekuléan and non-Kekuléan benzenoids are known (separately) for the formula in that particular position. For the significance of the full-drawn staircase line, see the text

formulas (n and s even) pertain to non-Kekuléans and/or Kekuléans. The computer-algorithm of Stojmenović et al. [17] is based on a boundary code, which enables the enumeration of all benzenoids with a given perimeter length, n_e [18,

Table 1. Numbers of benzenoid isomers from Stojmenović et al. [17, 20]*

h	n_i	d_s	C_nH_s	$|C_nH_s|$
15	8	5	$C_{54}H_{26}$	947291
	9	4	$C_{53}H_{25}$	395860
	10	3	$C_{52}H_{24}$	155656
	11	2	$C_{51}H_{23}$	55857
	12	1	$C_{50}H_{22}$	18396
16	10	4	$C_{56}H_{26}$	1132642
	11	3	$C_{55}H_{25}$	436698
	12	2	$C_{54}H_{24}$	156434
	13	1	$C_{53}H_{23}$	51691
	14	0	$C_{52}H_{22}$	16234
17	12	3	$C_{58}H_{26}$	1236839
	13	2	$C_{57}H_{25}$	440491
	14	1	$C_{56}H_{24}$	148430
	15	0	$C_{55}H_{23}$	46166
	16	−1	$C_{54}H_{22}$	13286
18	14	2	$C_{60}H_{26}$	1262442
	15	1	$C_{59}H_{25}$	424429
	16	0	$C_{58}H_{24}$	133713
	17	−1	$C_{57}H_{23}$	39143
	18	−2	$C_{56}H_{22}$	10587
19	16	1	$C_{62}H_{26}$	1223950
	17	0	$C_{61}H_{25}$	388180
	18	−1	$C_{60}H_{24}$	116648
	19	−2	$C_{59}H_{23}$	32042
	20	−3	$C_{58}H_{22}$	7885
20	18	0	$C_{64}H_{26}$	1140529
	19	−1	$C_{63}H_{25}$	345834
	20	−2	$C_{62}H_{24}$	97607
	21	−3	$C_{61}H_{23}$	25050
	22	−4	$C_{60}H_{22}$	5726
21	20	−1	$C_{66}H_{26}$	1029521
	21	−2	$C_{65}H_{25}$	296025
	22	−3	$C_{64}H_{24}$	79472
	23	−4	$C_{63}H_{23}$	18876
	24	−5	$C_{62}H_{22}$	3838
22	22	−2	$C_{68}H_{26}$	903415
	23	−3	$C_{67}H_{25}$	247089
	24	−4	$C_{66}H_{24}$	62027
	25	−5	$C_{65}H_{23}$	13652
	26	−6	$C_{64}H_{22}$	2467
23	24	−3	$C_{70}H_{26}$	771061
	25	−4	$C_{69}H_{25}$	200545
	26	−5	$C_{68}H_{24}$	47167
	27	−6	$C_{67}H_{23}$	9349
	28	−7	$C_{66}H_{22}$	1448
24	26	−4	$C_{72}H_{26}$	643859
	27	−5	$C_{71}H_{25}$	157989
	28	−6	$C_{70}H_{24}$	34324
	29	−7	$C_{69}H_{23}$	6124
25	28	−5	$C_{74}H_{26}$	522218
	29	−6	$C_{73}H_{25}$	120524
	30	−7	$C_{72}H_{24}$	24118
	31	−8	$C_{71}H_{23}$	3735

Table 1. (continued)

h	n_i	d_s	C_nH_s	$|C_nH_s|$
26	30	−6	$C_{76}H_{26}$	414305
	31	−7	$C_{75}H_{25}$	89288
	32	−8	$C_{74}H_{24}$	16114
	33	−9	$C_{73}H_{23}$	2146
27	32	−7	$C_{78}H_{26}$	319258
	33	−8	$C_{77}H_{25}$	63732
	34	−9	$C_{76}H_{24}$	10313
28	34	−8	$C_{80}H_{26}$	240218
	35	−9	$C_{79}H_{25}$	43833
	36	−10	$C_{78}H_{24}$	6214
29	36	−9	$C_{82}H_{26}$	174603
	37	−10	$C_{81}H_{25}$	29029
	38	−11	$C_{80}H_{24}$	3456
30	38	−10	$C_{84}H_{26}$	123790
	39	−11	$C_{83}H_{25}$	18198
31	40	−11	$C_{86}H_{26}$	84207
	41	−12	$C_{85}H_{25}$	10886
32	42	−12	$C_{88}H_{26}$	55451
	43	−13	$C_{87}H_{25}$	6042
33	44	−13	$C_{90}H_{26}$	34739
	45	−14	$C_{89}H_{25}$	3129
34	46	−14	$C_{92}H_{26}$	20811
35	48	−15	$C_{94}H_{26}$	11719
36	50	−16	$C_{96}H_{26}$	6155

* Supplementary to the data in a previous review [8].

19]. Their data [17], which correspond to the area to the left of the staircase-line in Fig. 2, cover all benzenoids with $n_e \leq 46$. All these data, linked explicitly to the appropriate C_nH_s formulas, are quoted in a previous review [20]. Most of them, for which the classification into Kekuléans and non-Kekuléans are known, are found in our main reference [8]. A supplement of unclassified numbers is given in Table 1. Out of these numbers those for odd-carbon atom formulas give, nevertheless, the numbers of non-Kekuléan benzenoid isomers (equal to the totals); hence the corresponding positions are marked with asterisks in Fig. 2; this is not the case for the even-carbon atom formulas.

Like the algorithm of Stojmenović et al. (see above) the computerized generations of benzenoids for $h \leq 10$ by the Düsseldorf−Zagreb group [7, 21, 22] take advantage of the fact that a benzenoid is uniquely determined by its perimeter. However, the number of hexagons, viz. h, is taken as the leading parameter (rather than n_e), making use of the addition of hexagons to the perimeter (cf. Sect. 2.2). The generated benzenoids with a given h value are classified according to their numbers of internal vertices (n_i), whereby an enumeration of the C_nH_s isomers is virtually achieved.

The complete data of the numbers of benzenoid isomers for $h = 13$ [8, 23] and $h = 14$ [8], along with the numbers of catacondensed benzenoids with $h = 15$ [8, 24], were deduced by means of an adaptation of a computational coding invented by the Düsseldorf−Zagreb group [25−29] and named DAST (dualist angle-restricted spanning tree). These data resulted in some asterisks in the upper part of Fig. 2 beyond (to the right of) the staircase-line. Hence they represent an extension of the data from Stojmenović et al. [17]. But also many of the unclassified numbers from the same source [17] were reproduced by an independent computation and classified so as to result in additional asterisks, now within (to the left of) the staircase-line of Fig. 2 [8, 23, 30].

The principles of the last mentioned computations were different from all of those mentioned above, although h was chosen as the leading parameter, and the additions of hexagons were employed. In these computations the different options [31] were exploited to a high degree, whereby the different types of additions (cf. Sect. 2.2) played an important role in the algorithm. The basic principles, which are of relevance to the computations in question, are treated in the next section.

Erratum: In Table 6 of Brunvoll et al. [8] the reference of footnote "a" is inappropriate. The corresponding values are published elsewhere [23], a paper which was in press when the mentioned chapter [8] was completed.

5 Building-Up and *Aufbau*

5.1 Theorems About the Building-Up of Benzenoids

It is well known (and obvious) that all benzenoids with $h + 1$ hexagons can be generated by additions of one hexagon each time in all the possible positions to all benzenoids with h hexagons. In this process, which sometimes is referred to as the building-up of benzenoids, there are the five possible types (i)−(v) of additions to the perimeter; cf. Sect. 2.2. Stronger statements, which are formulated below, have been proved.

Theorem: All benzenoids with $h + 1$ hexagons can be generated by (only) three types of additions to all benzenoids with h hexagons, viz. the (i) one-, (ii) two- and (iii) three-contact additions. All these three types are necessary.

The proof of the first part of this theorem is easy, but somewhat lengthy and therefore not included here.

The statement about necessity in the last part of the theorem is demonstrated by the sample of examples shown in Fig. 3. For the sake of illustration three examples are displayed in each of the cases (i), (ii) and (iii), the first one in each row being the smallest example. Only one example for each case would of course be sufficient to prove the last part of the theorem.

It is significant that (for $h > 1$) the L_1-, P_2- and L_3- mode hexagons are the only ones which possess free edges among the addition-mode hexagons. (Also an A_2-hexagon possesses a free edge, but that is not an addition mode.) These three

S. J. Cyvin, B. N. Cyvin, and J. Brunvoll

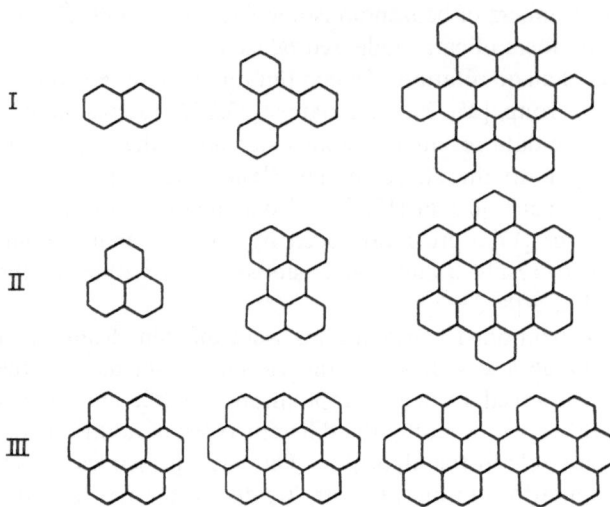

Fig. 3. Examples of benzenoids where the (**i**) one-, (**ii**) two- or (**iii**) three-contact addition is necessary when they are built up by means of these three addition types only

modes are exactly those acquired by the added hexagon in the cases (i), (ii) and (iii), respectively; cf. Sect. 2.2. In the examples of the three rows of Fig. 3 all the hexagons with free edges are exclusively in the modes (i) L_1, (ii) P_2 or (iii) L_3, respectively.

The opposite process of the building up is the tearing down of a benzenoid. The following statement is a direct corollary of the above theorem.

Corollary: It is possible to tear down any benzenoid completely (down to benzene) by removing one L_1-, P_2- or L_3-mode hexagon every time.

5.2 *Aufbau* Units

We shall refer to the generation of a class of benzenoid isomers (C_nH_s) from other classes of benzenoid isomers as *aufbau*. In this we follow Dias [32–34], who apparently transferred this term from the glossary which pertains to the Mendeleev periodic table for elements.

The three important types of addition (i), (ii) and (iii) have been described by Dias [16, 32–38] as an attachment of the unit (i) C_4H_2, (ii) C_3H or (iii) C_2, respectively. These are the *aufbau units* which Dias [32, 33] referred to as the *elementary* aufbau *units*.

The four-contact addition (iv) corresponds to an attachment of CH simultaneously with a loss in 2H; altogether $+C$, $-H$. For the five-contact addition (v) the "*aufbau* unit" is entirely negative, $-H_2$ [39, 40], if this designation can be used at all. Dias characterized the type (v) as a ring closure concomitant with an H_2 loss [40].

Dias [16, 32–34, 36–38] has also considered a type of an attachment of C_6H_2, which corresponds to a successive addition of two hexagons in P_2-modes

$(P_2 + P_2)$. Furthermore, he has considered another attachment of C_6H_2 [16, 32–35, 41], which also corresponds to a successive addition of two hexagons, but now as $L_3 + L_1$ [42].

5.3 *Aufbau* Principles

Dias [16], in his seminal paper, formulated a hypothesis about the *aufbau* of benzenoid isomers with proper reservations as to its general validity. The hypothesis was quoted elsewhere [7, 36, 38] without these reservations. In still more recent reviews Dias [32–34] quotes the hypothesis under the designation "*aufbau* principle". Nevertheless, it was demonstrated that this original *aufbau* principle of Dias, although it works in many special cases, is not generally valid. In contrast, we shall formulate other corollaries of the *Theorem* of Sect. 5.1 as general principles, which have been proved rigorously. The first part of the principles below is called the *fundamental* aufbau *principle*.

Principles: All benzenoid ($h > 1$) isomers with the formula C_nH_s can be generated by the following three types of additions (attachments) –
   (a) two-contact additions (C_3H attachments) to the $C_{n-3}H_{s-1}$ benzenoids,
   (b) one-contact additions (C_4H_2 attachments) to the $C_{n-4}H_{s-2}$ benzenoids, and
   (c) three-contact additions (C_2 attachments) to the $C_{n-2}H_s$ benzenoids.
   If one or two of the classes $C_{n-3}H_{s-1}$, $C_{n-4}H_{s-2}$ and $C_{n-2}H_s$ do not exist, then the remaining one or two classes are sufficient for the *aufbau*.

The fundamental *aufbau* principle is illustrated in Fig. 4. Here the *aufbau* units which come into operation are exactly the elementary *aufbau* units. The corresponding attachments of (i) C_4H_2, (ii) C_3H and (iii) C_2 are illustrated in Fig. 5 and pertain to the first example in each row (i)–(iii) of Fig. 4. But notice that the three types (a)–(c) specified in the *Principles* above correspond to the additions (ii), (i) and (iii) in that order. The formulation was chosen to be that way for several reasons, mainly because the type (a) is often sufficient for the *aufbau*, and because of our concern about extreme-left benzenoids [8] (see below). The last paragraph of the *Principles* gives rise to some general rules as specified in the following.

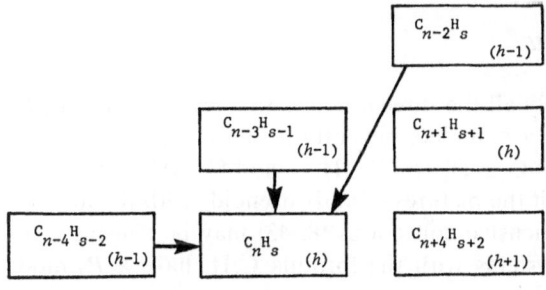

Fig. 4. Illustration of the fundamental *aufbau* principle

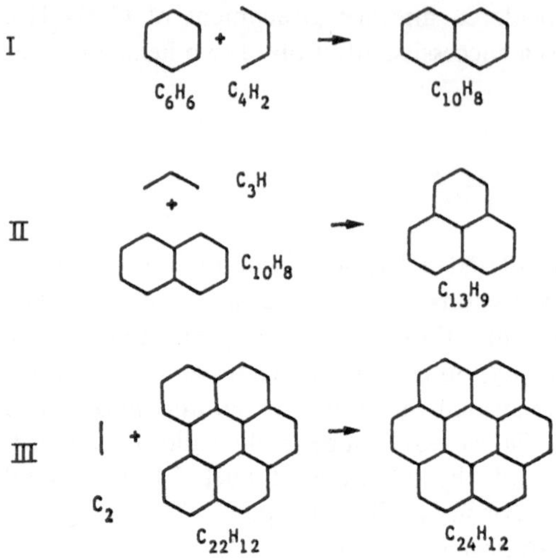

Fig. 5. The three types of attachments (i) C_4H_2, (ii) C_3H and (iii) C_2

The formulas in the first row of the periodic table for benzenoid hydrocarbons ($C_{4h+2}H_{2h+4}$) correspond to the catacondensed systems with $h > 1$. Since there are no other formulas above this row, all the $C_{4h+2}H_{2h+4}$ ($h > 1$) benzenoids are generated by the scheme (b) exclusively [16, 31].

The *extreme-left* benzenoids have by definition C_nH_s formulas at the extreme left of the rows of the periodic table [8]. Since these formulas have no formula to the left in the periodic table all the pericondensed extreme-left benzenoids (i.e. for $h > 2$) can be generated by the two schemes (a) and (c).

A *protrusive* benzenoid has a formula with nothing either above or to the left of it in the periodic table. Hence all the pericondensed protrusive benzenoids (i.e. for $h > 2$) can be generated by the scheme (c) exclusively [32].

Similar rules can be set for the formulas in the second row of the periodic table, viz. $C_{4h+1}H_{2h+3}$ ($h \geq 3$). The corresponding benzenoids are pericondensed with one internal vertex each. Firstly, since there are no formulas two rows above, it must be possible to generate the benzenoids in question by the schemes (a) and (b) above. Secondly, a stronger statement is true: all the $C_{4h+1}H_{2h+3}$ benzenoids can be generated (for $h > 3$) by the scheme (b) exclusively. The last statement is easily seen to be true since the systems in question consist of phenalene ($h = 3$) and phenalene with catacondensed appendage(s) (for $h > 3$).

5.4 Special Cases of *Aufbau*

It is not always necessary to apply all the three schemes (a)–(c) in the *Principles* of Sect. 5.3 in order to achieve a complete generation of the C_nH_s benzenoid isomers, even if all three formulas $C_{n-3}H_{s-1}$, $C_{n-4}H_{s-2}$ and $C_{n-2}H_s$ exist. These features are easily accounted for if the pictures of all benzenoids with the formula C_nH_s are available; two comprehensive collections [8, 43] may be consulted for numerous examples. If all the systems with the formula C_nH_s have a P_2-mode hexagon, then the scheme (a) is sufficient to generate them; if all of them have an L_1-mode hexagon, then the scheme (b) is sufficient; finally if they all have an L_3-mode hexagon, then (c) is sufficient. Furthermore, if all the systems of C_nH_s have either a P_2- or an L_1-mode hexagon (or both), then the two schemes (a) and (b) are sufficient, and so on.

Several of these properties are illustrated in Fig. 6. In many cases the scheme (a) alone is sufficient and is then indicated by a vertical arrow alone. That applies to most of the formulas for pericondensed benzenoids in Fig. 6. For $C_{24}H_{14}$, $C_{28}H_{14}$, $C_{31}H_{15}$, $C_{34}H_{16}$, $C_{36}H_{16}$ of Fig. 6 and $C_{41}H_{17}$ of Fig. 7 it is indicated, by one horizontal and one vertical arrow pointing to each box, that the schemes

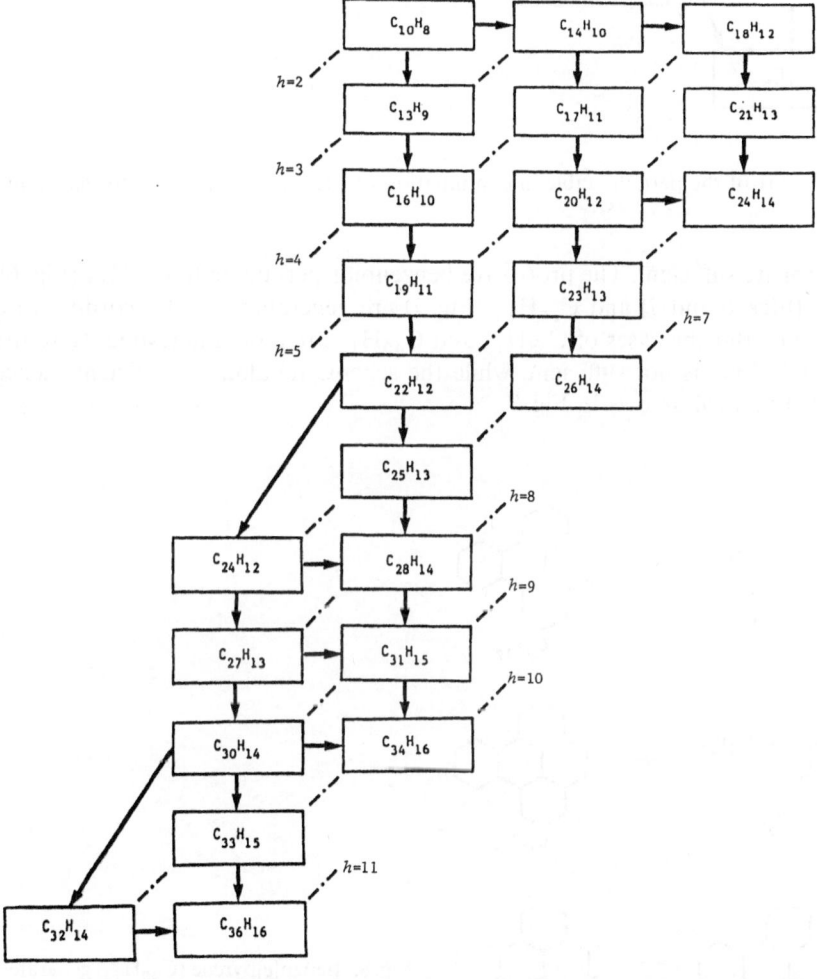

Fig. 6. A part of the periodic table for benzenoid hydrocarbons with the generation of the isomers indicated: *vertical arrow* corresponds to scheme (a), *horizontal arrow* to (b), and a *skew arrow* to the scheme (c) of *Principles* (Sect. 5.3)

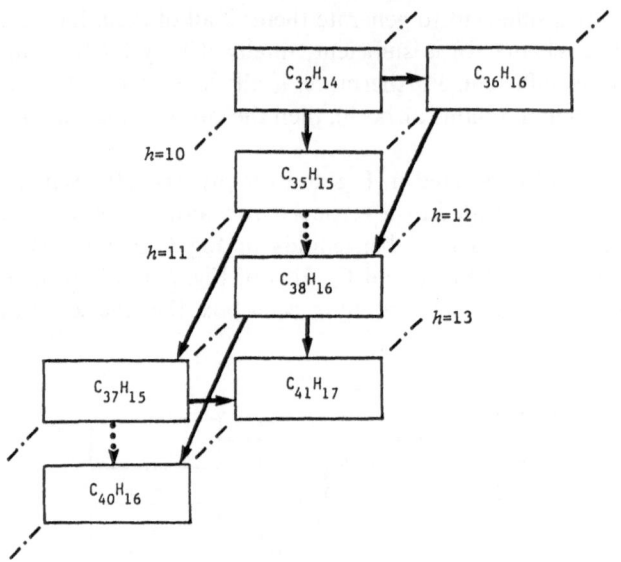

Fig. 7. A part of the periodic table in continuation of Fig. 6; see the legend to that figure

(a) and (b) are sufficient. The protrusive benzenoids pertaining to $C_{24}H_{12}$ (Fig. 6), $C_{32}H_{14}$ (Figs. 6 and 7) and $C_{37}H_{15}$ (Fig. 7) are generated by (c) according to a general rule. But the cases of $C_{38}H_{16}$ and $C_{40}H_{16}$ are more interesting. Here the scheme (a) alone is not sufficient, while the scheme (c) alone is sufficient; hence the dotted vertical arrows in Fig. 7.

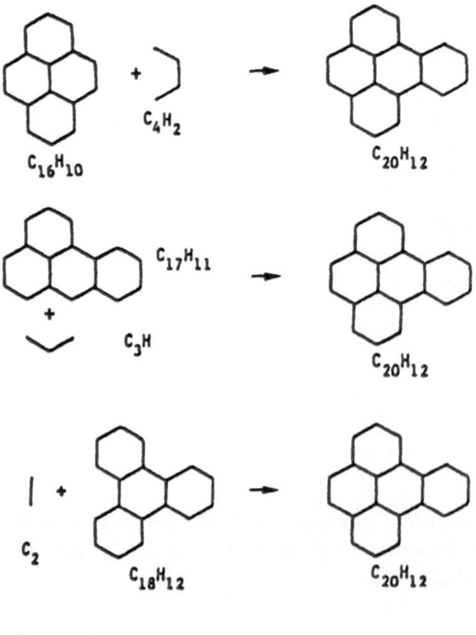

Fig. 8. Benzo[e]pyrene ($C_{20}H_{12}$) generated by the one-, two- and three-contact addition ($+C_4H_2$, $+C_3H$ and $+C_2$, respectively)

The original "*aufbau* principle" of Dias [16] (see also above) is not applicable to protrusive benzenoids. Apart from this, it has been formulated for the even-carbon atom systems, but can also be extended immediately to the odd-carbon atom systems. As such this incomplete principle works in most of the relevant cases of Figs. 6 and 7; only $C_{38}H_{16}$ and $C_{40}H_{16}$ are counterexamples therein.

It is no contradiction to the above statements that a benzenoid can often be generated by different schemes. The benzopyrenes $C_{20}H_{12}$ (benzo[a]- and benzo[e]pyrene), for instance, each contain an L_1-mode hexagon, P_2-mode hexagon(s) and L_3-mode hexagon(s). Thus they can be generated alternatively by either of the three schemes (a)–(c); see Fig. 8. This kind of multiple generations of a given benzenoid must be taken care of in algorithms for enumerations of classes of nonisomorphic benzenoids, not least in computer programming.

6 The Staircase-Like Boundary

6.1 Listing of Formulas

The C_nH_s formulas on the staircase-like boundary are those of most interest in the present chapter. Therefore an extensive listing of these formulas is given here; see Fig. 9. The values of the Dias parameter (d_s), which are indicated in the figure, are supposed to facilitate the identification of the positions of the different formulas in the periodic table for benzenoid hydrocarbons; the same is the case for the h values.

In Fig. 9 all the formulas pertain to extreme-left benzenoids except C_6H_6 (outside the periodic table) and four formulas at the start of the periodic table, viz. $C_{14}H_{10}$, $C_{17}H_{11}$, $C_{20}H_{12}$ and $C_{28}H_{14}$. The formulas for the smallest protrusive benzenoids are seen to be $C_{10}H_8$, $C_{24}H_{12}$, $C_{32}H_{14}$, $C_{37}H_{15}$, ... The staircase-like boundary (see Fig. 9) starts with a six-formula step followed by a four-formula step, while all the subsequent steps hold two or three formulas each. This property was stated previously without proof [8].

The keywords of Fig. 9, which are associated with the different symbols attached to most of the formulas, are discussed in subsequent sections.

6.2 Proving Some Propositions

Many important features of the staircase-like boundary can be deduced from the fundamental *aufbau* principle.

Propositions 1: (*a*) The staircase-like boundary has no one-formula step. (*b*) The staircase-like boundary has no plateau of more than one formula for pericondensed benzenoid(s).

In order to prove *Proposition 1* (*a*) consider the diagram below where each box represents a formula in the periodic table for benzenoid hydrocarbons. Suppose

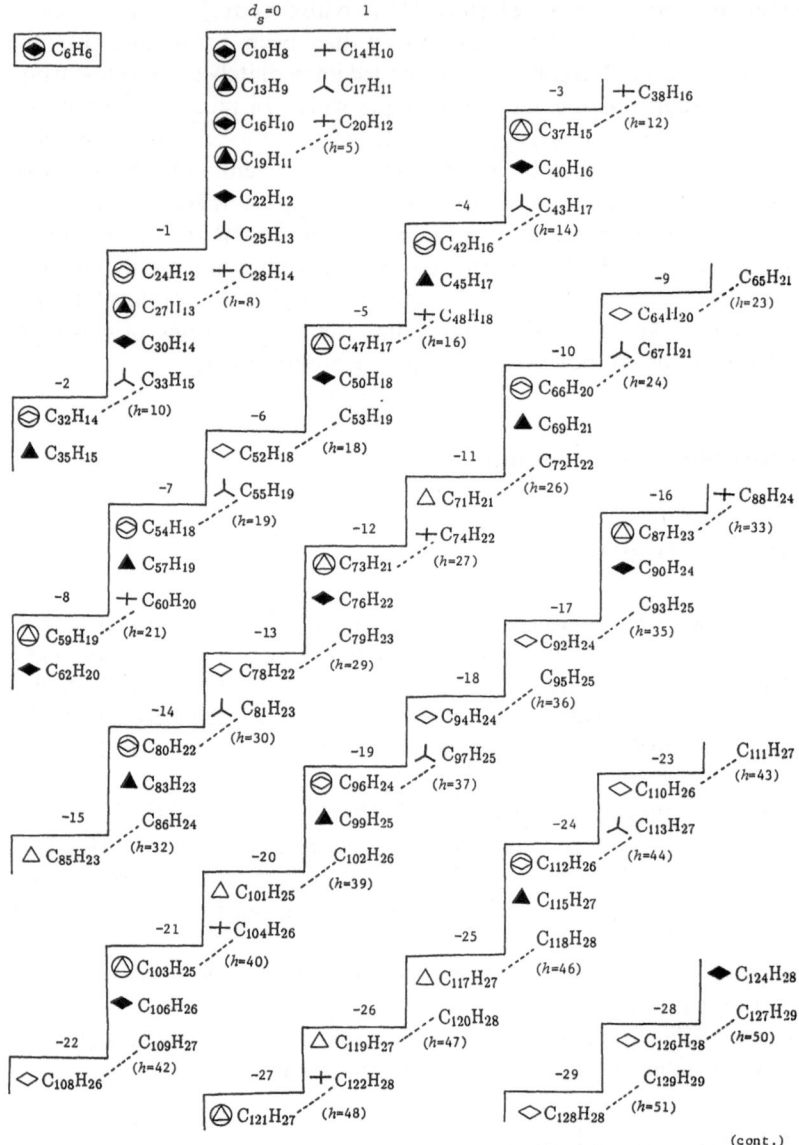

Fig. 9. Formulas ($C_n H_s$) on the staircase-like boundary of the periodic table for benzenoid hydrocarbons. The formula for benzene ($C_6 H_6$) is added

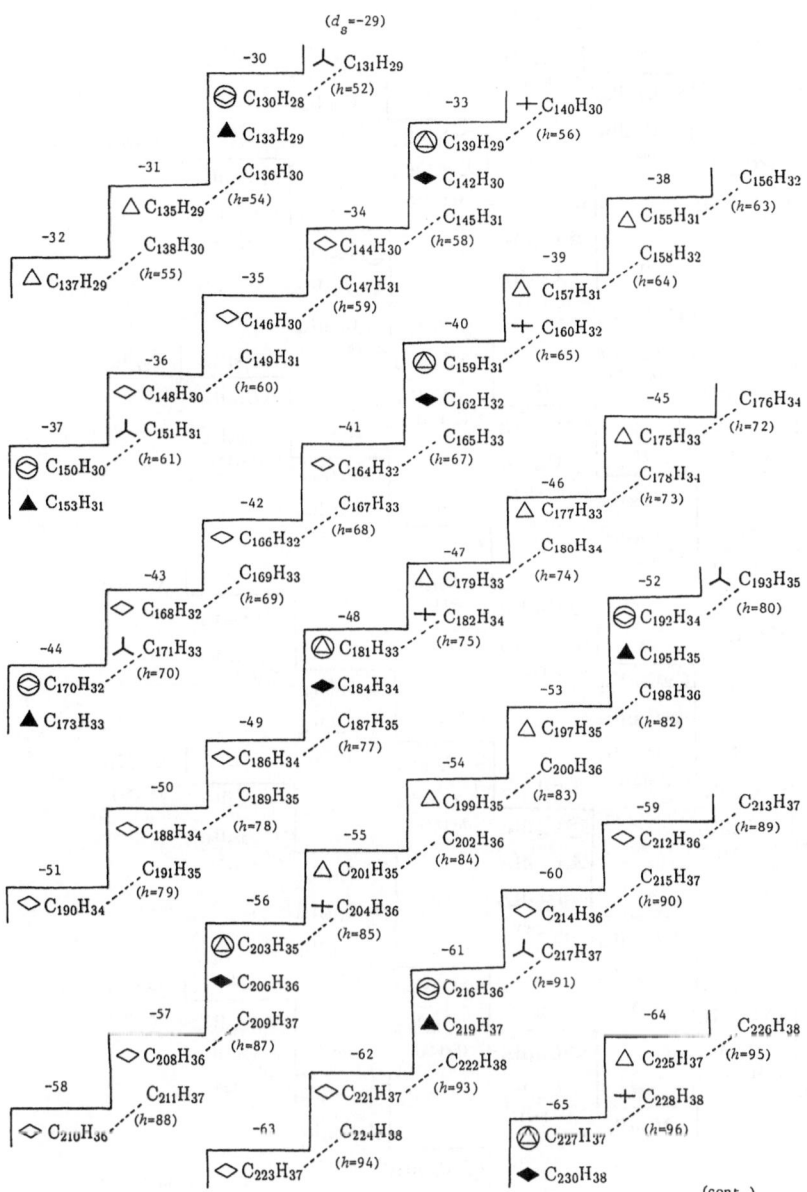

Fig. 9. (continued)

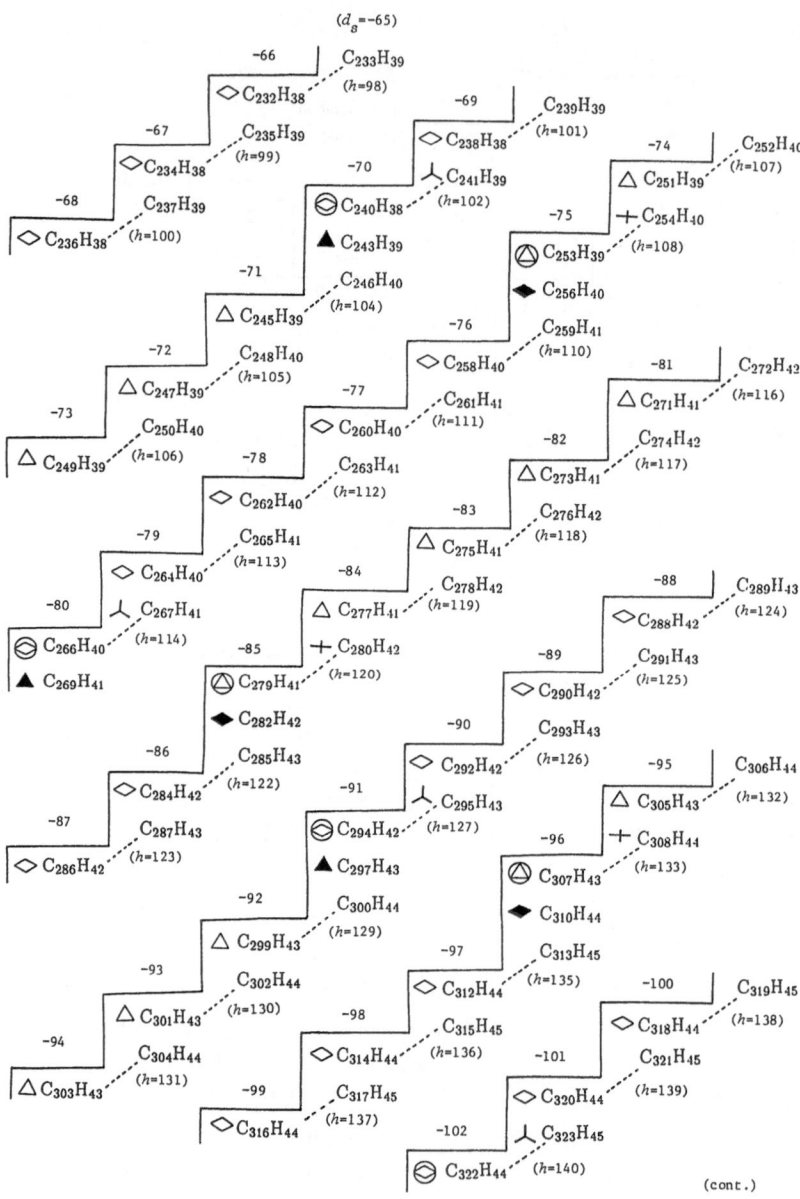

Fig. 9. (continued)

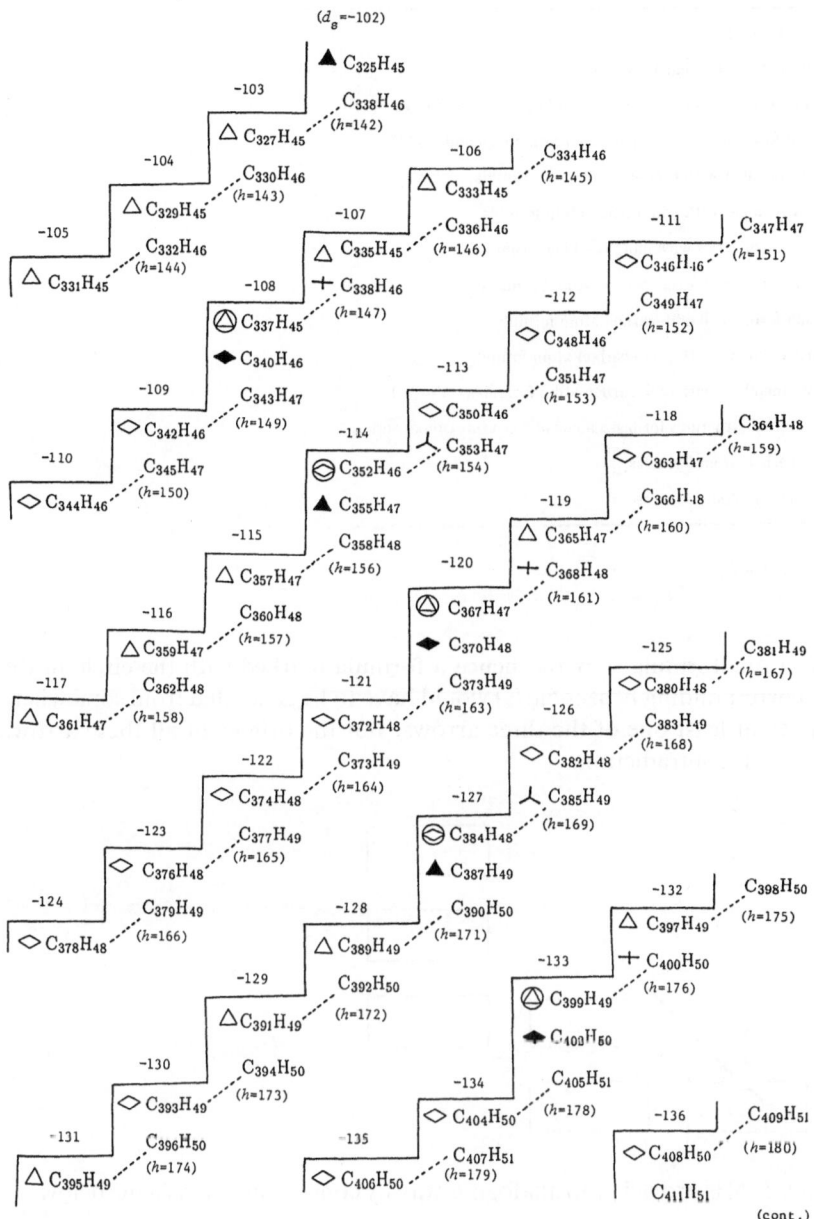

Fig. 9. (continued)

85

S. J. Cyvin, B. N. Cyvin, and J. Brunvoll

Explanation of symbols —

(1) one–isomer series (circular benzenoids):

⬣ ground form with even–carbon atom formula (n and s even);

▲ ground form with odd–carbon atom formula (n and s odd);

⬡ higher member with even–carbon atom formula;

⬡ higher member with odd–carbon atom formula

(2) constant–isomer series with more than one isomer:

◆ ground forms with even–carbon atom formulas;

▲ ground forms with odd–carbon atom formulas;

◇ higher members with even–carbon atom formulas;

△ higher members with odd–carbon atom formulas

(3) excised internal structures for benzenoids of constant–isomer series:

✛ even–carbon atom formulas;

⅄ odd–carbon atom formulas.

Fig. 9. (continued)

there was a one-formula step and hence a formula marked with the circle in the box. The corresponding benzenoid(s) would have to be generated from somewhere according to at least one of the three arrows. But the origins of all these arrows are empty — a contradiction.

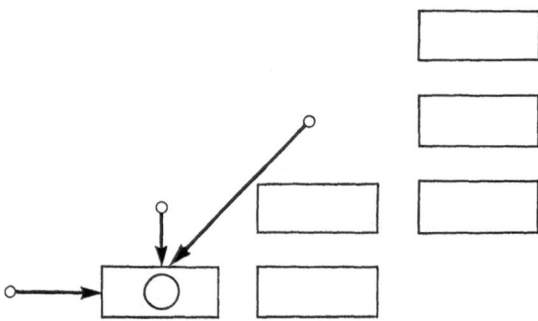

Proposition 1 (b) is proved in an analogous way by considering the scheme below.

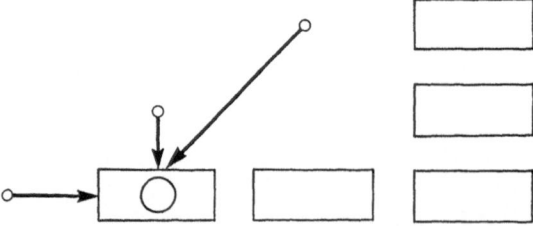

In the previous chapter [8] it was stated that extremal benzenoids have no coves and no fjords. Below we shall prove two propositions which, when taken together represent a slightly more precise statement. But first we recall the definition of an extremal benzenoid [8]. An *extremal* benzenoid has the maximum number of internal vertices (n_i) for a given number of hexagons (h). The appropriate formulas C_nH_s represent exclusively one or more extremal benzenoids each. These formulas are found on the staircase-like boundary of the periodic table at the ends of each diagonal line going through formulas for the same h. From Fig. 9 one may easily identify the formulas for the smallest extreme-left benzenoids which are not extremal; they are $C_{25}H_{13}$, $C_{33}H_{15}$, $C_{38}H_{16}$, $C_{43}H_{17}$, ... On the other hand, all extremal benzenoids for $h > 1$ are also extreme-left.

Propositions 2: (*a*) An extremal benzenoid has no cove and no fjord. (*b*) An extreme-left benzenoid has no fjord.

Suppose that C_nH_s is the formula for an extreme-left benzenoid with a cove. By covering the cove (in the language of Section 2.2) a new benzenoid is created with one hexagon more and three more internal vertices. It has the formula $C_{n+1}H_{s-1}$ which is situated in the periodic table relatively to C_nH_s as indicated in the below scheme. At the same time the scheme shows the only possible shape of the staircase-like boundary in the vicinity of C_nH_s and $C_{n+1}H_{s-1}$, as can easily be deduced from *Propositions 1* (*a*) and *1* (*b*). It is found that C_nH_s is not compatible with an extremal benzenoid. That proves *Proposition 2* (*a*).

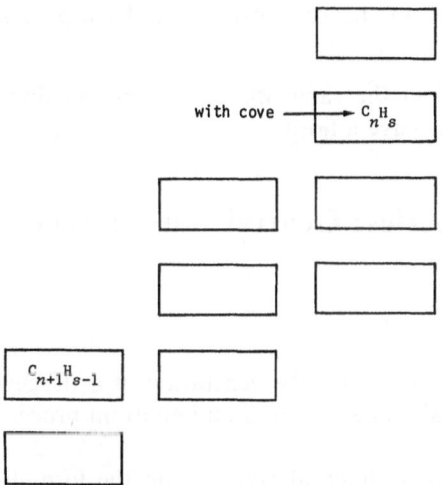

As a result of the above proof it may be stated: An extreme-left benzenoid with a cove is not an extremal benzenoid.

In order to prove *Proposition 2* (*b*) suppose that C_nH_s is the formula for an extreme-left benzenoid with a fjord. By immersing a hexagon in the fjord a new benzenoid is created with four more internal vertices. The position of the new formula, viz. $C_{n-2}H_s$, is indicated below. Now it is impossible to construct a staircase-like boundary between the two formulas without violating *Proposition 1* (*a*), *1* (*b*) or both — a contradiction.

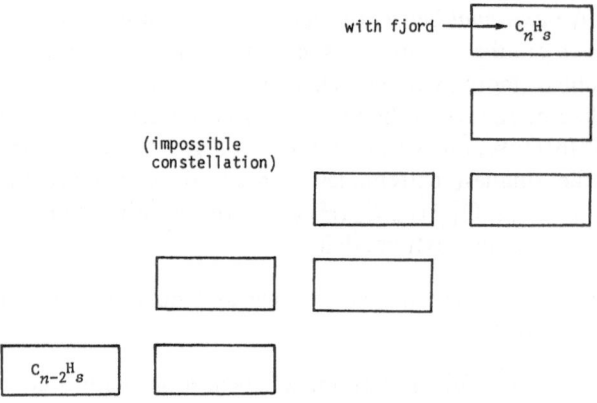

Proposition 3: A protrusive benzenoid with the formula on top of a three-formula step has no bay.

This proposition is proved straightforwardly in an analogous way to *Proposition 2* (*a*) or (*b*); the positions of the formulas C_nH_s and $C_{n+2}H_s$ are considered. A protrusive benzenoid is also extremal and has consequently neither a cove or a fjord.

Proposition 4: An extreme-left benzenoid in the row just above the top of a three-formula step has no cove.

This proposition can again be proved in the same way as the propositions above. The benzenoids in question do not have a fjord.

7 Constant-Isomer Benzenoid Series: General Introduction

7.1 Circumscribing

It has been pointed out (with documentation) that the generation of a (larger) benzenoid by circumscribing another (smaller) benzenoid is an important process in the studies of benzenoid isomers [8].

Let B be a benzenoid with h hexagons, n_i internal vertices and the formula C_nH_s, and assume that it can be circumscribed. Further, let $B' \equiv$ circum-B have the formula $C_{n'}H_{s'}$, and assume (h', n_i') to be the pair of its invariants which correspond to (h, n_i). Different connections between the invariants of B and B' have been given previously [8, 20]. Some more of them are found below, as can easily be deduced from the fundamental and simple relations:

$$h' - h = s, \qquad n_i' - n = 0 \tag{7}$$

One has

$$h' = 3h - n_i + 4 = (1/2)(n + s) + 1 \tag{8}$$

and for the coordinates (d'_s, n'_i) of $\mathbf{B}'$:

$$d'_s = -h = (1/2)(s - n) - 1 \tag{9}$$

$$n'_i = 4h - n_i + 2 = n \tag{10}$$

The last relation (10) implies the obvious fact that the number of internal vertices of $\mathbf{B}'$ equals the total numbers of vertices of $\mathbf{B}$. One has furthermore

$$(n'; s') = (8h - 3n_i + 16; 2h - n_i + 10)$$
$$= (n + 2s + 6; s + 6) \tag{11}$$

Assume now that $\mathbf{B}$ can be circumscribed k times. Then the resulting benzenoid, k-circum-$\mathbf{B}$, has the formula given by:

$$(n_k; s_k) = (6k^2 + 4(k + 1)h - (2k + 1)n_i + 2(4k + 1);$$
$$6k + 2h - n_i + 4)$$
$$= (6k^2 + 2ks + n; 6k + s) \tag{12}$$

It is easily perceived that the following proposition must be valid.

Proposition 5: Any circumscribed benzenoid has all its free edges in L_3-mode hexagons.

As a corollary of this proposition such a benzenoid can only be generated by scheme (c) out of the three schemes in *Principles* of Sect. 5.3.

7.2 Historical

The interesting phenomenon of *constant-isomer series* of benzenoids was discovered by Dias when he, as long ago as his seminal paper [16], pointed out that the formulas $C_{24}H_{12}$, $C_{54}H_{18}$, $C_{96}H_{24}$, $C_{150}H_{30}$, ... correspond to one isomer each, viz. coronene, circumcoronene, dicircumcoronene, tricircumcoronene, ... In our notation (cf. Sect. 2.1):

$$|C_{24}H_{12}| = |C_{54}H_{18}| = |C_{96}H_{24}| = |C_{150}H_{30}| = 1 \tag{13}$$

This sequence of formulas is an example of a *one-isomer series* of benzenoids. Soon after it had been reported the other five one-isomer series were detected [37, 39, 45] and are collected elsewhere [20, 43, 46]. The first constant-isomer series of benzenoids with formulas representing more than one isomer each was again detected by Dias [45]. We write

$$|C_{22}H_{12}| = |C_{52}H_{18}| = |C_{94}H_{24}| = |C_{148}H_{30}| = 3 \tag{14}$$

Here $C_{22}H_{12}$ comprises the three isomers anthanthrene, benzo[*ghi*]perylene and triangulene, the forms of which have been displayed many times; they are also found in the previous review [8] (see also below). The subsequent formulas of (14) correspond to successive circumscribing of the three benzenoids under consideration.

7.3 Ground Forms and Higher Members

A constant-isomer series of benzenoids is characterized by the following relation for the cardinalities.

$$|n; s| = |6k^2 + 2ks + n; s + 6k| \tag{15}$$

cf. Eq. (12). The relation (15) should be obeyed for all $k = 0, 1, 2, \ldots$. The formula (C_nH_s) for $k = 0$ is assumed to correspond to the set of the smallest benzenoids belonging to the series. They are referred to as the *ground forms* [20, 43]. The set consists, of course, of one ground form in the case of one-isomer series. The larger benzenoids obtained by circumscribing the ground forms shall presently be referred to as *higher members*.

How can we know whether a benzenoid with the formula C_nH_s is a member (ground form or higher member) of a one-isomer series? The relation

$$|n; s| = 1 \tag{16}$$

is a necessary and sufficient condition, provided that $C_nH_s \neq C_{17}H_{11}$.

8 Circular Benzenoids and One-Isomer Benzenoid Series

8.1 Circular Benzenoids

A *circular* benzenoid [8] is defined by having the largest number of hexagons (h_{max}) for a given perimeter length or s value. It has been found [15, 47]

$$h_{max} = \lfloor (1/12) (s^2 - 6s + 12) \rfloor = 1 + Q,$$
$$Q = \lfloor (1/12) (s^2 - 6s) \rfloor \tag{17}$$

where the floor function is employed: $\lfloor x \rfloor$ is the largest integer smaller than or equal to x. Then the formulas for the circular benzenoids, say O, are given by

O: $(s + 2Q; s)$ $\hspace{2cm}$ (a)

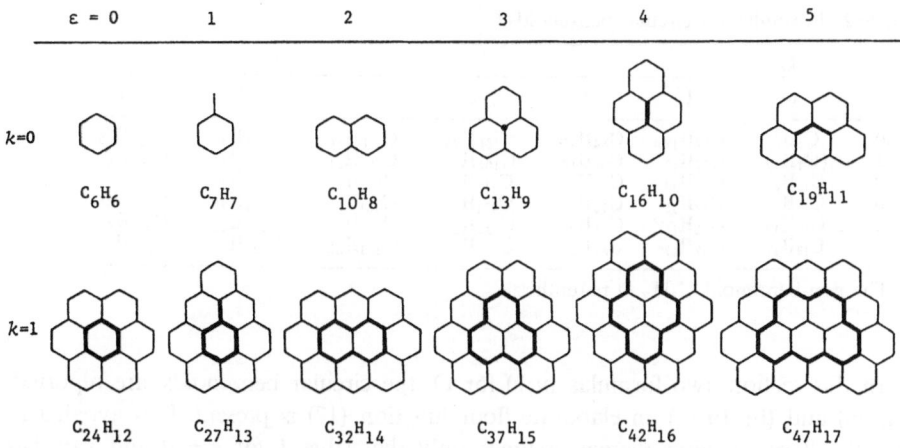

$k=0$

C_6H_6 C_7H_7 $C_{10}H_8$ $C_{13}H_9$ $C_{16}H_{10}$ $C_{19}H_{11}$

$k=1$

$C_{24}H_{12}$ $C_{27}H_{13}$ $C_{32}H_{14}$ $C_{37}H_{15}$ $C_{42}H_{16}$ $C_{47}H_{17}$

Fig. 10. The smallest circular benzenoids with the non-benzenoid C_7H_7 included

where Q is given by Eq. (17), and s may assume all the possible values for this parameter: $s = 6, 8, 9, 10, 11, \ldots$.

Each formula for a circular benzenoid in Fig. 9 is marked by a circle (with some other symbol inside it). It can be shown that, apart from C_6H_6 and the appropriate formulas in the two first (abnormally high) steps in the periodic table, the circular benzenoids always have their formulas at the top of three-formula steps.

There is exactly one circular benzenoid for each formula of O. Furthermore, these benzenoids exhibit six characteristic shapes to be accounted for in the following. These shapes were at least known to Balaban [48], who investigated their important role in the studies of annulenes; cf. also some later works in this connection [15, 49].

Figure 10 shows the forms of the smallest circular benzenoids. For the sake of systematization the non-benzenoid system for C_7H_7 is included. All larger systems are obtained by circumscribing those which are displayed. In Fig. 10, the second row ($k = 1$) is obtained by circumscribing the systems in the first row ($k = 0$). The different shapes (in the columns of Fig. 10) are identified by the labels $\varepsilon = 0, 1, 2, 3, 4, 5$.

The validity of the following proposition is easily perceived by looking at Fig. 10.

Proposition 6: Any circular benzenoid with $h = 7$ and $h \geq 10$ has all its free edges in the six L_3-mode hexagons which it possesses.

This is a special case of *Proposition 5* in Sect. 7.1.

With the aid of the previous analyses of annulenes [15, 48, 49] the following alternative expression was found for the formulas of the circular benzenoids in general.

O: $\qquad (6k^2 + 12k + (2k + 3)\,\varepsilon + 4 + 2\lfloor 1 - (\varepsilon/6)\rfloor; 6k + \varepsilon + 6)$ $\qquad$ (b)

In Table 2 the formulas of quite a few of the first (smallest) circular benzenoids are listed.

Table 2. Formulas for circular benzenoids*

ϵ	k						
	0	1	2	3	4	5	6
0	C_6H_6	$C_{24}H_{12}$	$C_{54}H_{18}$	$C_{96}H_{24}$	$C_{150}H_{30}$	$C_{216}H_{36}$	$C_{294}H_{42}$
1	(C_7H_7)	$C_{27}H_{13}$	$C_{59}H_{19}$	$C_{103}H_{25}$	$C_{159}H_{31}$	$C_{227}H_{37}$	$C_{307}H_{43}$
2	$C_{10}H_8$	$C_{32}H_{14}$	$C_{66}H_{20}$	$C_{112}H_{26}$	$C_{170}H_{32}$	$C_{240}H_{38}$	$C_{322}H_{44}$
3	$C_{13}H_9$	$C_{37}H_{15}$	$C_{73}H_{21}$	$C_{121}H_{27}$	$C_{181}H_{33}$	$C_{253}H_{39}$	$C_{337}H_{45}$
4	$C_{16}H_{10}$	$C_{42}H_{16}$	$C_{80}H_{22}$	$C_{130}H_{28}$	$C_{192}H_{34}$	$C_{266}H_{40}$	$C_{352}H_{46}$
5	$C_{19}H_{11}$	$C_{47}H_{17}$	$C_{87}H_{23}$	$C_{139}H_{29}$	$C_{203}H_{35}$	$C_{279}H_{41}$	$C_{367}H_{47}$

* The non-benzenoid C_7H_7 in parentheses

In this section, two formulas $(n; s)$ for O, the circular benzenoids, are reported, viz. (a) and (b). In (a) an elaborate floor function (17) is present. It is avoided in (b); the floor function therein assumes only the value 1 for $\varepsilon = 0$ and vanishes for $\varepsilon > 0$. But instead, the form (b) contains two parameters (k, ε), while (a) contains only one, viz. s. We shall refer to the two formulas for O as being in (a) the *Harary-Harborth picture* and (b) the *Balaban picture*, respectively. It is recalled that they originate from the analysis of Harary and Harborth [44] in the case of (a) and from Balaban [48] in the case of (b). Under this scope all the inequalities and deductions from them in the previous chapter [8] are in the Harary-Harborth picture.

8.2 One-Isomer Benzenoid Series

The circular benzenoids are members of the six one-isomer series of benzenoids which exist and correspond to the six characteristic shapes (cf. Fig. 10). The ground forms occur for $k = 0$ when $\varepsilon \neq 1$ and $k = 1$ when $\varepsilon = 1$; their formulas are specifically: C_6H_6, $C_{10}H_8$, $C_{13}H_9$, $C_{16}H_{10}$, $C_{19}H_{11}$ and $C_{27}H_{13}$. The first higher members in each series, which emerge from circumscribing, are $C_{24}H_{12}$, $C_{32}H_{14}$, $C_{37}H_{15}$, $C_{42}H_{16}$, $C_{47}H_{17}$ and $C_{59}H_{19}$.

In Fig. 9 the ground forms are indicated by black symbols (in the present case inscribed in circles): rhombs for even-carbon atom formulas, triangles for odd-carbon atom formulas. Higher members of the one-isomer series are identified by circles with a white rhomb or a white triangle for the even- and odd-carbon atom formulas, respectively.

For the one-constant benzenoid isomer series which starts with C_6H_6 the general formula $(6(k + 1)^2; 6(k + 1))$ [20] is found readily from Eq. (12). A number of additional formulas of this kind have been given elsewhere [20, 34]. In the above formula $k = 0$ pertains to benzene (C_6H_6), while $k > 0$ reproduce coronene and (poly)circumscribed coronene(s). The formula of Eq. (4), viz. $C_{600}H_{60}$, emerges as the special case for $k = 9$: octacircumcoronene or "nonacircumbenzene".

In the following section we report a substantially more general formulation; completely general expressions in the two pictures (of Harary-Harborth and of Balaban) are given, in fact, for the formulas of constant-isomer series, both for the ground forms and the higher members.

As a preparation to the general formulations we shall transfer the two parameters (k, ε) for the ground forms of one-isomer series into a new pair, written in braces as $\{j, \delta\}$. Now the formulas in question after C_6H_6 are labelled by $j = 0$ and $\delta = 1, 2, 3, 4, 6$ for $C_{10}H_8$, $C_{13}H_9$, $C_{16}H_{10}$, $C_{19}H_{11}$ and $C_{27}H_{13}$, respectively. The δ value which was skipped, viz. $\delta = 5$, belongs to C_6H_6, but with $j = -1$.

9 Constant-Isomer Benzenoid Series: General Formulations

9.1 Introductory Remarks

It was proved that there is no cove and no fjord among the extremal benzenoids. This is a necessary condition for the feature that all such benzenoids can be circumscribed. In general, the absence of coves and fjords is not a sufficient condition for this feature [8]. Nevertheless it is inferred that all extremal benzenoids, say A, can be circumscribed and not only once, but an infinite number of times. We go still a step further. Let $(n; s)$ be the formula of A and $(n'; s')$ of circum-A as in Eq. (11). If all the $(n; s)$ benzenoids A are circumscribed, it is inferred that the created circum-A benzenoids represent all the $(n'; s')$ isomers. Then there is obviously the same number of the A and circum-A systems. The properties described in this paragraph are very plausible and represent the foundation of the algorithms [46, 50, 51] and general formulations [51] for constant-isomer benzenoid series, which have been put forward. Also some attempts to provide rigorous proofs are in progress.

Let again the formulas $(n; s)$ and $(n'; s')$ pertain to A and circum-A, respectively, where A is an extremal benzenoid. Then, by virtue of *Proposition 5* (Sect. 7.1) it should be clear that circum-A must be protrusive (and hence also extremal).

The above description aims at the following conclusion. All extremal benzenoids are members of constant-isomer series. The ground forms for $h > 2$ have formulas just below those of the circular benzenoids in the periodic table. Therefore, apart from C_6H_6 and the abundant occurrence of the ground form formulas in the two first steps of the staircase-like boundary of the periodic table for benzenoids, they are constantly found in the middle of the three-formula steps. The formulas for higher members are without exception protrusive benzenoids. The reader may follow the description of this paragraph by observing the symbols in Fig. 9: black rhombs and triangles indicate the even- and odd-carbon atom formulas, respectively, for the ground forms. White rhombs and triangles indicate the even- and odd-carbon formulas for higher members. In both cases the encircled symbols pertain to one-constant isomer series.

9.2 Harary-Harborth Picture

From the formula apparatus in our main reference [8], founded on the Harary-Harborth [44] relations, one obtains the following equation for the perimeter length (n_e) of an extremal benzenoid.

$$(n_e/2)_{\min} = \lceil (12h - 3)^{1/2} \rceil = R \tag{18}$$

93

The invariant n_e, also equal to the number of external vertices, assumes its minimum given by (18) for a given h. Here the ceiling function is employed: $\lceil x \rceil$ is the smallest integer larger than or equal to x. Now the formula for an extremal benzenoid, A, reads in general [8]

A: $\qquad (2h + 1 + R; 3 + R)$ $\qquad\qquad$ (c)

where R is given by (18), and $h = 1, 2, 3, 4, \ldots$. Correspondingly for circumscribed A one has [8, 23]

circum-A: $\quad (2h + 13 + 3R; 9 + R)$ $\qquad\qquad$ (d)

In order to get the formulas $(n; s)$ exclusively for the ground forms of constant-isomer benzenoid series one could use expression (c) for A and insert $h = 1, 2, 3, 4, \ldots$, but skip the formulas which coincide with circum-A (d). This occurs for the numbers of hexagons

$$h' = h + 3 + R \qquad\qquad (19)$$

viz. $h' = 7, 10, 12, 14, 16, 18, 19, \ldots$ (an irregular sequence; cf. Fig. 9).

Instead of the elaborate specification of the above paragraph we are able to present an explicit general formula for the ground forms, $(n; s)$, when switching from h to s as the leading parameter. Start from the formula (a) for a circular benzenoid, O, as given in Sect. 8.1. Increase the first coefficient (number of carbon atoms) by three and the second (number of hydrogens) by one. This corresponds to the C_3H attachment (or two-contact addition), which brings us to the formula just below, as it should be. Finally replace $s + 1$ by s. In this way the following general expression for the formulas of the ground forms, say G, is achieved [51].

G: $\qquad (n; s) = (s + 2\lfloor (1/12) (s^2 - 8s + 19) \rfloor; s)$ $\qquad\qquad$ (20)

Here $s = 6, 8, 9, 10, 11, \ldots$ as above (in Sect. 8.1). The formulas for the ground forms of the one-constant isomer series (cf. Sect. 8.2) are reproduced by $s = 6, 8, 9, 10, 11$ and 13, while $s = 12$, which comes in-between, pertains to the constant-isomer series (14) with three isomers.

The formula (20) can readily be generalized to all members of constant-isomer series by means of Eq. (12). One obtains [51]:

$$(n_k; s_k) = (6k^2 + (2k + 1) s + 2\lfloor (1/12) (s^2 - 8s + 19) \rfloor; s + 6k) \quad (21)$$

Here $k = 0$ pertains to the ground forms, $k > 0$ to higher members.

9.3 Balaban Picture

The general expression for the formulas of the ground forms of constant-isomer benzenoid series in the Balaban picture was worked out from the formula (b) of Sect. 8.1. The new pair of parameters $\{j, \delta\}$ was introduced to be consistent with the description in Sect. 8.2. The net result reads [51]

$$G: \qquad (n; s) = (6j^2 + 12j + (2j + 3)\,\delta + 7 + 2\lfloor \delta/6 \rfloor; 6j + \delta + 7) \qquad (22)$$

Here, in lexicographic order, corresponding to increasing s values in Eq. (20), $j = 0, 1, 2, 3, \ldots$ and $\delta = 1, 2, 3, 4, 5, 6$. As such Eq. (22) does not cover the formula C_6H_6 ($s = 6$; benzene). In order to include that, the $\{j, \delta\}$ pair $\{0, 1\}$ should be preceded by $\{-1, 5\}$. In general the pair of parameters in $\{j, \delta\}$ for a given s may be determined by

$$j = \lfloor (1/6)\,(s - 8) \rfloor \qquad (23)$$

and

$$\delta = s - 6j - 7 = s - 1 - 6\lfloor (1/6)\,(s - 2) \rfloor \qquad (24)$$

In Table 3 a number of formulas for the ground forms of constant-isomer benzenoid series are listed.

Table 3. Formulas for ground forms of constant-isomer benzenoid series*

δ	j						
	0	1	2	3	4	5	6
1	$C_{10}H_8$	$C_{30}H_{14}$	$C_{62}H_{20}$	$C_{106}H_{26}$	$C_{162}H_{32}$	$C_{230}H_{38}$	$C_{310}H_{44}$
2	$C_{13}H_9$	$C_{35}H_{15}$	$C_{69}H_{21}$	$C_{115}H_{27}$	$C_{173}H_{33}$	$C_{243}H_{39}$	$C_{325}H_{45}$
3	$C_{16}H_{10}$	$C_{40}H_{16}$	$C_{76}H_{22}$	$C_{124}H_{28}$	$C_{184}H_{34}$	$C_{256}H_{40}$	$C_{340}H_{46}$
4	$C_{19}H_{11}$	$C_{45}H_{17}$	$C_{83}H_{23}$	$C_{133}H_{29}$	$C_{195}H_{35}$	$C_{269}H_{41}$	$C_{355}H_{47}$
5	$C_{22}H_{12}$	$C_{50}H_{18}$	$C_{90}H_{24}$	$C_{142}H_{30}$	$C_{206}H_{36}$	$C_{282}H_{42}$	$C_{370}H_{48}$
6	$C_{27}H_{13}$	$C_{57}H_{19}$	$C_{99}H_{25}$	$C_{153}H_{31}$	$C_{219}H_{37}$	$C_{297}H_{43}$	$C_{387}H_{49}$

* They are preceded by C_6H_6 for $\{j, \delta\} = \{-1, 5\}$.

Again, by means of Eq. (12) it was deduced [51]:

$$(n_k; s_k) = (6(j + k)^2 + 2(j + k)(\delta + 6) + 2k + 3\delta + 7 + 2\lfloor \delta/6 \rfloor;$$
$$6(j + k) + \delta + 7) \qquad (25)$$

9.4 Three-Parameter Code

As a result of the treatment in the Balaban picture each formula for a member of a constant-isomer series (ground form or a higher member) is identified by a triple, say $\{j, \delta, k\}$. Here $k = 0$ corresponds to the ground forms; we write

Table 4. Numbers of isomers for constant-isomer series with even-carbon atom formulas

j	δ	Ground form h Formula	Higher member formulas	Kek.	Non—Kek.	Total
−1	5	1 C_6H_6	$C_{24}H_{12}, C_{54}H_{18}, C_{96}H_{24}$, $C_{150}H_{30}, C_{216}H_{36}, C_{294}H_{42}$, $C_{384}H_{48}$, ...	1[a]	0	1[a]
0	1	2 $C_{10}H_8$	$C_{32}H_{14}, C_{66}H_{20}, C_{112}H_{26}$ $C_{170}H_{32}, C_{240}H_{38}, C_{322}H_{44}$, ...	1[a]	0	1[a]
0	3	4 $C_{16}H_{10}$	$C_{42}H_{16}, C_{80}H_{22}, C_{130}H_{28}$, $C_{192}H_{34}, C_{266}H_{40}, C_{352}H_{46}$, ...	1[a]	0	1[a]
0	5	6 $C_{22}H_{12}$	$C_{52}H_{18}, C_{94}H_{24}, C_{148}H_{30}$, $C_{214}H_{36}, C_{292}H_{42}, C_{382}H_{48}$, ...	2[b]	1[b]	3[b]
1	1	9 $C_{30}H_{14}$	$C_{64}H_{20}, C_{110}H_{26}, C_{168}H_{32}$, $C_{238}H_{38}, C_{314}H_{44}, C_{408}H_{50}$, ...	3[b]	1[b]	4[b]
1	3	13 $C_{40}H_{16}$	$C_{78}H_{22}, C_{128}H_{28}, C_{190}H_{34}$, $C_{264}H_{40}, C_{350}H_{46}$, ...	3[c,d]	1[e]	4[f]
1	5	17 $C_{50}H_{18}$	$C_{92}H_{24}, C_{146}H_{30}, C_{212}H_{36}$, $C_{290}H_{42}, C_{380}H_{48}$, ...	7[c,d]	2[c]	9[c]
2	1	22 $C_{62}H_{20}$	$C_{108}H_{26}, C_{166}H_{32}, C_{236}H_{38}$, $C_{318}H_{44}, C_{412}H_{50}$, ...	12[c]	4[g-k]	16[f]
2	3	28 $C_{76}H_{22}$	$C_{126}H_{28}, C_{188}H_{34}, C_{262}H_{40}$, $C_{348}H_{46}$, ...	12[g-l]	4[g-k]	16[f]
2	5	34 $C_{90}H_{24}$	$C_{144}H_{30}, C_{210}H_{36}, C_{288}H_{42}$, $C_{378}H_{48}$, ...	27[g-k]	12[g-k]	39[f]
3	1	41 $C_{106}H_{26}$	$C_{164}H_{32}, C_{234}H_{38}, C_{316}H_{44}$, $C_{410}H_{50}$, ...	38[g-l]	19[g-k]	57[f]
3	3	49 $C_{124}H_{28}$	$C_{186}H_{34}, C_{260}H_{40}, C_{346}H_{46}$, ...	38[i,j]	19[i,j]	57[i,j]
3	5	57 $C_{142}H_{30}$	$C_{208}H_{36}, C_{286}H_{42}, C_{376}H_{48}$, ...	86[j]	47[j]	133[i,j]
4	1	66 $C_{162}H_{32}$	$C_{232}H_{38}, C_{314}H_{44}, C_{408}H_{50}$, ...	128[j]	71[j]	199[i,j]
4	3	76 $C_{184}H_{34}$	$C_{258}H_{40}, C_{344}H_{46}$, ...	128[j]	71[j]	199[i,j]
4	5	86 $C_{206}H_{36}$	$C_{284}H_{42}, C_{374}H_{48}$, ...	264[m]	164[m]	428[i,j]
5	1	97 $C_{230}H_{38}$	$C_{312}H_{44}, C_{406}H_{50}$, ...	373	243	616[i,j]
5	3	109 $C_{256}H_{40}$	$C_{342}H_{46}$, ...	373	243	616[i,j]
5	5	121 $C_{282}H_{42}$	$C_{372}H_{48}$, ...	749	516	1265[i,j]
6	1	134 $C_{310}H_{44}$	$C_{404}H_{50}$, ...	1055	745	1800[n]
6	3	148 $C_{340}H_{46}$	....	1055	745	1800[n]
6	5	162 $C_{370}H_{48}$	....	2022	1517	3539[n]
7	1	177 $C_{402}H_{50}$	....	2765	2132	4897[n]
7	3	193 $C_{436}H_{52}$	....	2765	2132	4897

$\{j, \delta, 0\} \equiv \{j, \delta\}$. In this notation, for instance, the six ground forms and five higher members of the one-constant isomer series (cf. Sect. 8.2), which are depicted in Fig. 10, are labelled in the following way when arranged in the same way as the forms of the figure.

$$\{-1, 5, 0\} \qquad - \qquad \{0, 1, 0\} \quad \{0, 2, 0\} \quad \{0, 3, 0\} \quad \{0, 4, 0\}$$
$$\{-1, 5, 1\} \quad \{0, 6, 0\} \quad \{0, 1, 1\} \quad \{0, 2, 1\} \quad \{0, 3, 1\} \quad \{0, 4, 1\}$$

10 Constant-Isomer Benzenoid Series: Preliminary Numerical Results, and Depictions

10.1 Numbers

The known numbers of isomers for the constant-isomer series of benzenoids are listed in Tables 4 and 5 for the even- and odd-carbon atom formulas, respectively. In the former case (Table 4) the separate numbers for Kekuléan (abbreviated Kek.) and non-Kekuléan (abbr. Non-Kek.) systems are given. This listing goes far beyond the mapping with asterisks in Fig. 2, but moves only on the staircase-like boundary (cf. the formulas with black symbols in Fig. 9).

Most of the data under consideration are taken from literature. The references already cited above need two supplements [52, 53].

10.2 Forms

It is useful to have some concrete pictures of the ground forms for constant-isomer benzenoid series, in continuation of Fig. 10 for the one-isomer series. Some of them are displayed in our main reference [8] and more of them elsewhere [20, 43], the latter citation [43] containing detailed documentations to previous works. All the complete sets from these sources are collected in Fig. 11 (without too much overlap with the previous chapter [8]).

The Kekuléan systems belonging to constant-isomer benzenoid series are invariably normal (marked **n** in Fig. 11) and have $\Delta = 0$. The non-Kekuléans are marked **o** with the Δ value indicated in a subscript (o_Δ). In Fig. 11 also the excised internal structures [8] (see also below) are indicated by contours in bold. The same is found in Fig. 10 and subsequent figures.

◀ (Notes to Table 4)

[a] Elk (1980) [52]; [b] Dias (1982) [16]; [c] Dias (1984) [45]; [d] Dias (1984) [39]; [e] Dias (1986) [37]; [f] Stojmenović, Tošić and Doroslovački (1986) [17]; [g] Dias (1990) [33]; [h] Dias (1990) [46]; [i] Dias (1990) [50]; [j] Dias (1990) [53]; [k] Brunvoll and Cyvin (1990) [20]; [l] Dias (1990) [34]; [m] Brunvoll, Cyvin and Cyvin (1991) [8]; [n] Cyvin and Brunvoll (1991) [51].

Table 5. Numbers of isomers for constant-isomer series with odd-carbon atom formulas

j	δ	Ground forms h	Formula	Higher member formulas	Total (Non–Kek.)
0	2	3	$C_{13}H_9$	$C_{37}H_{15}, C_{73}H_{21}, C_{121}H_{27}, C_{181}H_{33}, C_{253}H_{39}, C_{337}H_{45}, ...$	1[a]
0	4	5	$C_{19}H_{11}$	$C_{47}H_{17}, C_{87}H_{23}, C_{139}H_{29}, C_{203}H_{35}, C_{279}H_{41}, C_{367}H_{47}, ...$	1[a]
0	6	8	$C_{27}H_{13}$	$C_{59}H_{19}, C_{103}H_{25}, C_{159}H_{31}, C_{227}H_{37}, C_{307}H_{43}, C_{399}H_{49}, ...$	1[b]
1	2	11	$C_{35}H_{15}$	$C_{71}H_{21}, C_{119}H_{27}, C_{179}H_{33}, C_{251}H_{39}, C_{335}H_{45}, ...$	2[c,d]
1	4	15	$C_{45}H_{17}$	$C_{85}H_{23}, C_{137}H_{29}, C_{201}H_{35}, C_{277}H_{41}, C_{365}H_{47}, ...$	4[d]
1	6	20	$C_{57}H_{19}$	$C_{101}H_{25}, C_{157}H_{31}, C_{225}H_{37}, C_{305}H_{43}, C_{397}H_{49}, ...$	4[c,d]
2	2	25	$C_{69}H_{21}$	$C_{117}H_{27}, C_{177}H_{33}, C_{249}H_{39}, C_{333}H_{45}, ...$	13[d]
2	4	31	$C_{83}H_{23}$	$C_{135}H_{29}, C_{199}H_{35}, C_{275}H_{41}, C_{363}H_{47}, ...$	20[d]
2	6	38	$C_{99}H_{25}$	$C_{155}H_{31}, C_{223}H_{37}, C_{303}H_{43}, C_{395}H_{49}, ...$	20[d]
3	2	45	$C_{115}H_{27}$	$C_{175}H_{33}, C_{247}H_{39}, C_{331}H_{45}, ...$	48[e-g]
3	4	53	$C_{133}H_{29}$	$C_{197}H_{35}, C_{273}H_{41}, C_{361}H_{47}, ...$	74[e-g]
3	6	62	$C_{153}H_{31}$	$C_{221}H_{37}, C_{301}H_{43}, C_{393}H_{49}, ...$	74[e-g]
4	2	71	$C_{173}H_{33}$	$C_{245}H_{39}, C_{329}H_{45}, ...$	174[f,g]
4	4	81	$C_{195}H_{35}$	$C_{271}H_{41}, C_{359}H_{47}, ...$	258[f,g]
4	6	92	$C_{219}H_{37}$	$C_{299}H_{43}, C_{391}H_{49}, ...$	258[f,g]
5	2	103	$C_{243}H_{39}$	$C_{327}H_{45}, ...$	550[f,g]
5	4	115	$C_{269}H_{41}$	$C_{357}H_{47}, ...$	796[f,g]
5	6	128	$C_{297}H_{43}$	$C_{389}H_{49}, ...$	796[f,g]
6	2	141	$C_{325}H_{45}$	....	1634[h]
6	4	155	$C_{355}H_{47}$	....	2302[h]
6	6	170	$C_{387}H_{49}$	....	2302[h]

[a] Elk (1980) [52]; [b] Knop, Szymanski, Jeričević and Trinajstić (1983) [21]; [c] Dias (1986) [37]; [d] Stojmenović, Tošić and Doroslovački (1986) [17]; [e] Dias (1990) [46]; [f] Dias (1990) [50]; [g] Dias (1990) [53]; [h] Cyvin and Brunvoll (1991) [51].

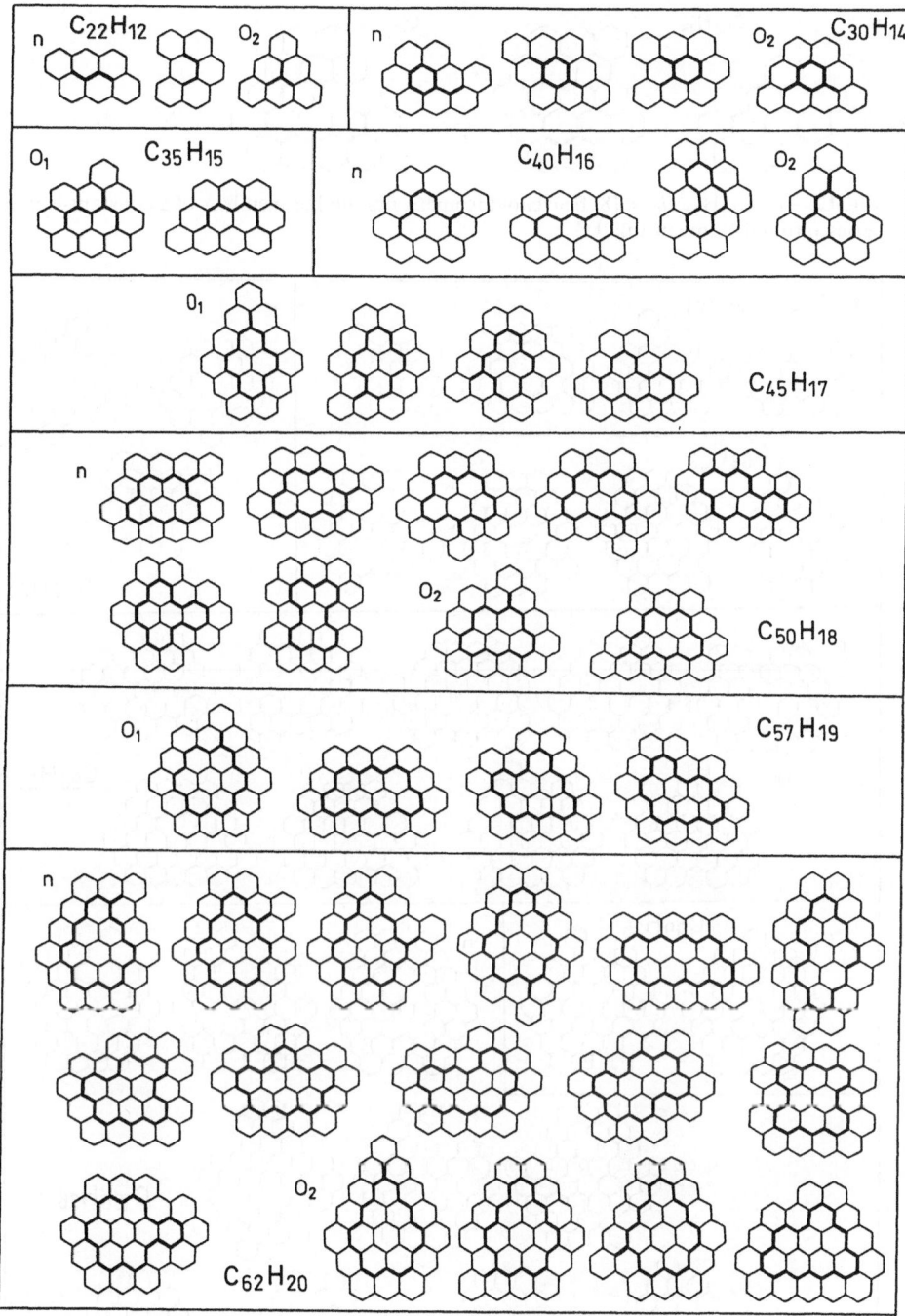

Fig. 11. The ground forms for constant-isomer series of benzenoids ($h = 6$ and $9 \leq h \leq 22$)

S. J. Cyvin, B. N. Cyvin, and J. Brunvoll

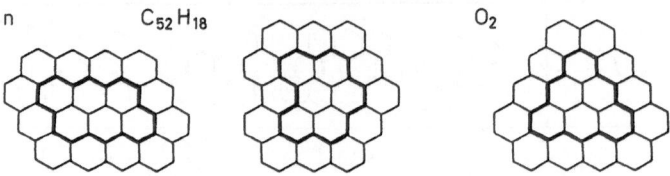

n $C_{52}H_{18}$ O_2

Fig. 12. The $C_{52}H_{18}$ ($h = 18$) benzenoid isomers, first higher members of a constant-isomer series (ground forms in Fig. 11)

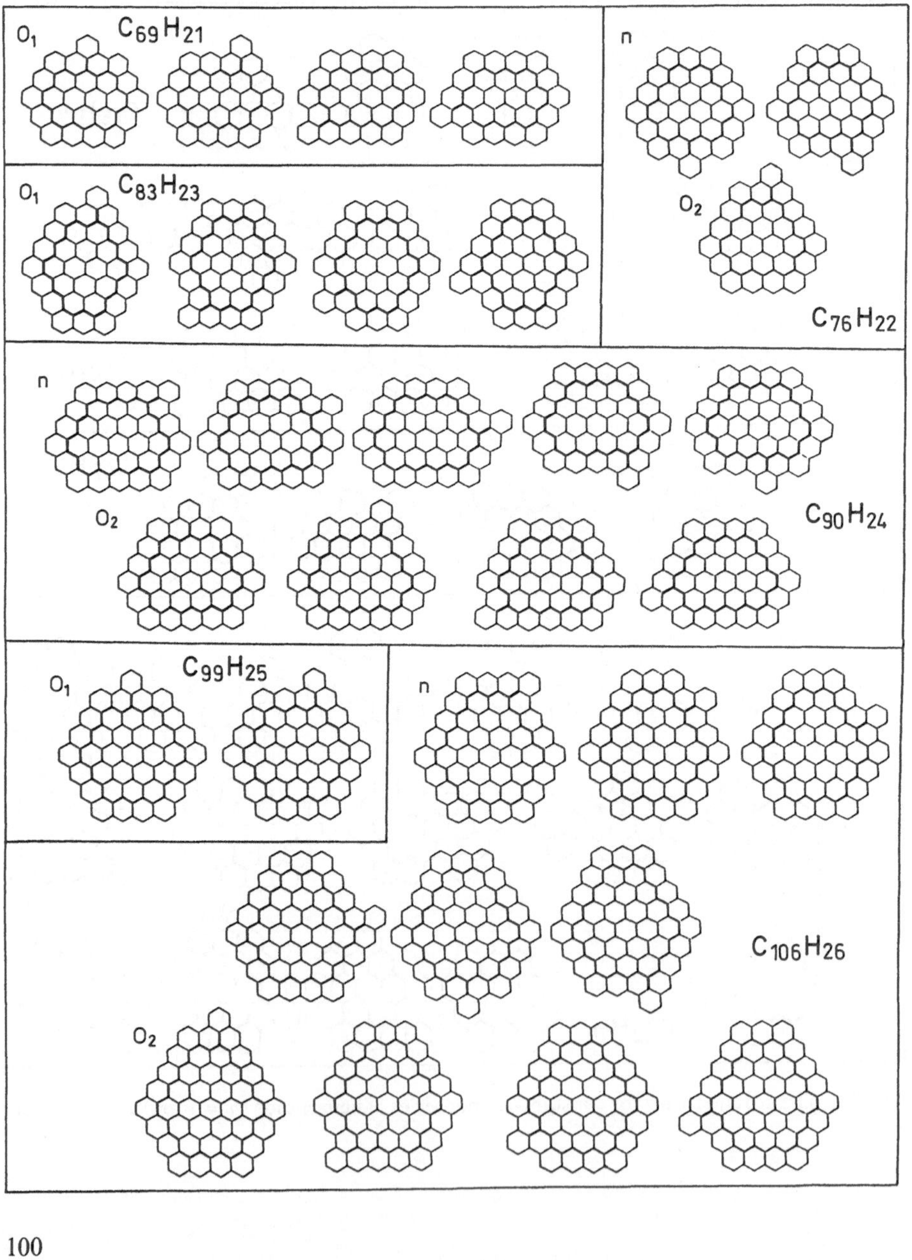

O_1 $C_{69}H_{21}$ n

O_1 $C_{83}H_{23}$

O_2

$C_{76}H_{22}$

n

O_2 $C_{90}H_{24}$

$C_{99}H_{25}$

O_1 n

$C_{106}H_{26}$

O_2

The first three benzenoids in Fig. 11 pertain to the first formula ($C_{22}H_{12}$) in Eq. (14). The next formula ($C_{52}H_{12}$) is accounted for by the first higher members of this series; see Fig. 12. These three instructive forms are also found in one of the references cited above [43]. Moreover, all three of them or the two Kekuléan ones, have been displayed several times by Dias [32–34, 36–38, 46, 53–55].

In Fig. 13 a part of the ground forms for the following constant-isomer benzenoid series are displayed: (1) $C_{69}H_{21}$ ($h = 25$), (2) $C_{76}H_{22}$ ($h = 28$), (3) $C_{83}H_{23}$ ($h = 31$), (4) $C_{90}H_{24}$ ($h = 34$), (5) $C_{99}H_{25}$ ($h = 38$), (6) $C_{106}H_{26}$ ($h = 41$). They should be supplemented with the respective circumscribed benzenoids of the sets of isomers as specified in the following.

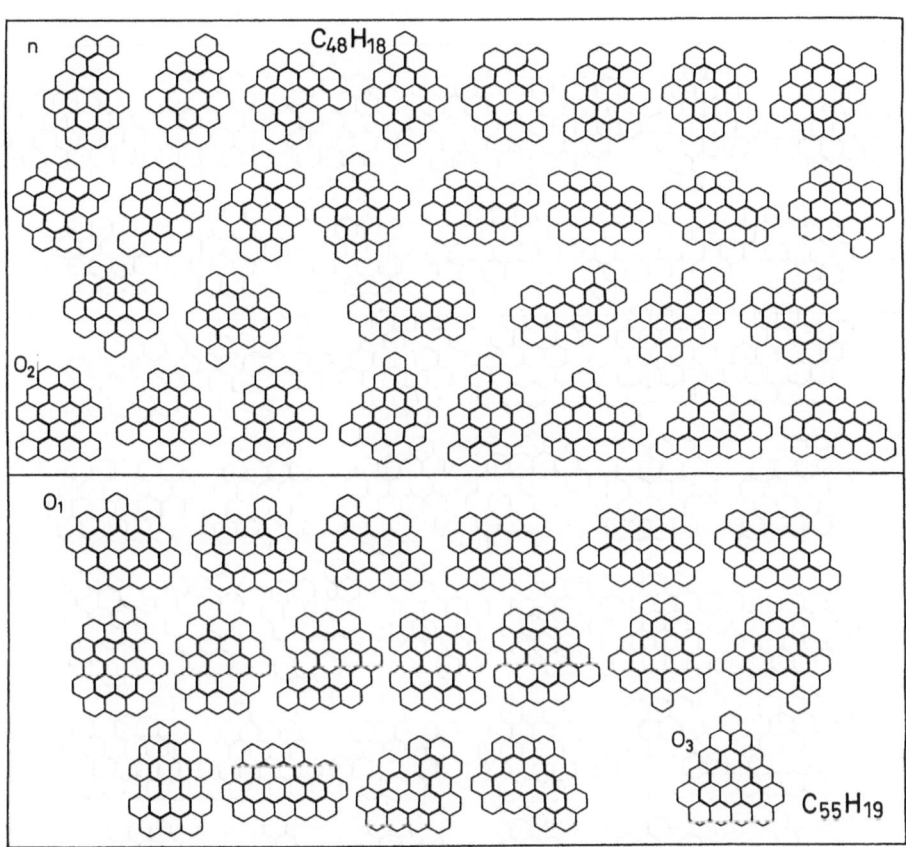

Fig. 14. The 30 $C_{48}H_{18}$ ($h = 16$) and 18 $C_{55}H_{19}$ ($h = 19$) benzenoid isomers

◄

Fig. 13. Parts of the ground forms for constant-isomer series of benzenoids ($25 \leq h \leq 41$); they should be supplemented with circumscribed benzenoids according to the text

(1) The 9 $C_{33}H_{15}$ ($h = 10$) systems $- 8o_1 + o_3$ [8];

(2) 13 $C_{38}H_{16}$ ($h = 12$) $- 10n + 3o_2$ [8];

(3) 16 $o_{43}H_{16}$ ($h = 14$) $- 15o_1 + o_3$ [8];

(4) 30 $C_{48}H_{18}$ ($h = 16$) $- 22n + 8o_2$ $-$ see Fig. 14;

(5) 18 $C_{55}H_{19}$ ($h = 19$) $- 17o_1 + o_3$ $-$ see Fig. 14;

(6) 47 $C_{60}H_{20}$ ($h = 21$) $- 32n + 15o_2$ $-$ see Fig. 15.

Fig. 15. The 47 $C_{60}H_{20}$ ($h = 21$) benzenoid isomers

11 Special *Aufbau* Procedure

A special *aufbau* procedure emerges by a modification of the fundamental *aufbau* principle of Sect. 5.3. The points (a) and (b) are retained, but the last point (c) is replaced by

(c′) circumscribing the $C_{n-2s+6}H_{s-6}$ systems .

The above formula is consistent with the inversion of the relations in (11) as

$$(n; s) = (n' - 2s' + 6; s' - 6) \tag{26}$$

As we shall see, the procedure is not generally valid for generating all isomers for given formulas.

Assume now that a benzenoid characterized by the invariants (h, n_i) and having the formula $(n; s)$ is circumscribed. The process implies, as is seen from Eq. (7), an addition of s hexagons. The number of internal vertices increases by $2s - 6$. The circumscribing of $C_n H_s$ is also properly described as an attachment of $C_{2s+6}H_6$. Examples of circumscribing are found in Fig. 10.

In the present procedure the schemes (a) and (b) may lead to isomorphic systems, of course, but the generation according to (c′) is exclusive. This fact is apparent from *Proposition 5* and its corollary (Sect. 7.1). But how can we know if the circumscribing can generate all the nonisomorphic benzenoids which are not obtained from (a) and (b)? As a matter of fact this is not always the case. Firstly, it is not guaranteed that all the systems in question are generated if only the $C_{n-2s+6}H_{s-6}$ benzenoids are circumscribed (if they exist). Sometimes it is necessary to include non-benzenoids, biphenyl $C_{12}H_{10}$ being the smallest example. Therefore the formulation under (c′) says "systems" rather than "benzenoids". The systems to be circumscribed may even consist of disconnected parts. Secondly, it may happen that benzenoids not generated by (a) and (b) can not be obtained by circumscribing at all. The above complications will not be encountered in the following application of the special *aufbau* procedure to ground forms of constant-isomer series for benzenoids.

The mentioned complications may cause serious difficulties under an application of the present *aufbau* procedure in general. But anyhow it is a great advantage that the systems under (c′) are considerably smaller than those under (c) in the fundamental *aufbau* principle.

12 Application of the Special *Aufbau* Procedure to Ground Forms of Constant-Isomer Benzenoid Series

12.1 Introductory Remarks

Let again G represent the ground forms of a constant-isomer series for benzenoids, of which the formula is $C_n H_s$ or $(n; s)$. In the following treatment the expression for G (22) in the Balaban picture (Sect. 9.3) is invoked. Also $(n; s) \equiv \{j, \delta\}$.

Since each G is an extreme-left benzenoid (and even extremal) only the two schemes (a) and (c′) of the special *aufbau* procedure (cf. Sect. 11) are of interest here. The cardinality of C_nH_s (for $h > 2$) can accordingly be split into two parts as is symbolized by:

$$|C_nH_s| = |C_nH_s(a)| + |C_nH_s(c')| \tag{27}$$

At the beginning of the periodic table the scheme (a) is sufficient for a complete *aufbau* of the ground forms, which means that the second part in (27) vanishes. It was found by inspection and is seen from Figs. 6 and 7 that this occurs for the G formulas $C_{13}H_9, C_{16}H_{10}, C_{19}H_{11}, C_{22}H_{12}, C_{27}H_{13}, C_{30}H_{14}$ and $C_{35}H_{15}$. The first case when the cardinality can not be obtained from the scheme (a) alone, occurs for $C_{40}H_{16}$. Here the *aufbau* is accomplished by (a) two-contact additions to $C_{37}H_{15}$ circumphenalene (see Fig. 10), which results in two systems, and (c′) circumscribings of the $C_{14}H_{10}$ benzenoids, viz. the two $h = 3$ catacondensed systems (anthracene and phenanthrene).

The method which is outlined above is nothing else than a part of an algorithm due to Dias [46, 50].

12.2 First Part of the Cardinality

In order to inspect the first term on the right-hand side of Eq. (27) closer we ask ourselves first on which circular benzenoid the two-contact addition according to (a) should be executed. It is clear that the formula of this circular benzenoid is obtained by subtracting C_3H from C_nH_s. In terms of the three-parameter code it was found that:

$$(n - 3; s - 1) \equiv \{0 - \lfloor \delta/6 \rfloor, \delta - 1 + 6\lfloor (7 - \delta)/6 \rfloor,$$
$$j + \lfloor \delta/6 \rfloor - \lfloor (7 - \delta)/6 \rfloor\} \tag{28}$$

This somewhat awkward formula is a straightforward consequence of Eq. (25). Let us introduce three new parameters so that (28) transforms into the simpler form

$$(n - 3; s - 1) \equiv \{o, \delta', j'\} \tag{29}$$

For the sake of brevity we shall demonstrate the relation (28) only under "normal" behaviour, viz. for $\delta = 2, 3, 4$ and 5. In these instances one has $\lfloor \delta/6 \rfloor = \lfloor (7 - \delta)/6 \rfloor = 0$, and therefore

$$o = 0, \quad \delta' = \delta - 1, \quad j' = j \quad (\delta = 2, 3, 4, 5) \tag{30}$$

Hence it is inferred by (28) that

$$(n - 3; s - 1) \equiv \{0, \delta - 1, j\} \quad (\delta = 2, 3, 4, 5) \tag{31}$$

Table 6. Three-parameter codes for $(n; s)$ and $(n - 3; s - 1)$, where $(n; s)$ represents the ground forms of a constant-isomer benzenoid series

$(n; s) \equiv \{j, \delta, 0\}$	$(n-3; s-1) \equiv \{o, \delta', j'\}$
$\{j, 1, 0\}$	$\{0, 6, j-1\}$
$\{j, 2, 0\}$	$\{0, 1, j\}$
$\{j, 3, 0\}$	$\{0, 2, j\}$
$\{j, 4, 0\}$	$\{0, 3, j\}$
$\{j, 5, 0\}$	$\{0, 4, j\}$
$\{j, 6, 0\}$	$\{-1, 5, j+1\}$

On inserting the three parameters from (31) as $\{j, \delta, k\}$ in (25) we obtain

$$\{0, \delta - 1, j\}$$
$$\equiv (6j^2 + 2j(\delta + 5) + 2j + 3(\delta - 1) + 7; 6j + \delta + 6)$$
$$= (6j^2 + 12j + (2j + 3)\delta + 4; 6j + \delta + 6) \tag{32}$$

But this is exactly the expression obtained from (25) with $k = 0$, or directly from Eq. (22), on substracting three units from n and one unit from s. Table 6 gives a full account of the three-parameter codes for $(n - 3; s - 1)$.

When we have identified the circular benzenoid $\{o, \delta', j'\}$, which should be subjected to the two-contact additions in order to derive $|C_nH_s(a)|$, for which $(n; s) \equiv \{j, \delta\}$, we introduce the following notation for this number.

$$|C_nH_s(a)| = P(o, \delta', j') \tag{33}$$

It is a simple combinatorial problem to derive the P functions of (33). The whole Fig. 13 in fact consists of an abundance of examples. Below we will show the derivation of an explicit expression for P in only one case, viz. $o = 0$, $\delta' = 2$. These are the systems $C_{13}H_9$, $C_{37}H_{15}$, $C_{73}H_{21}$, $C_{121}H_{27}$ ... (cf. Table 5), from which the ground forms or part of the ground forms $C_{16}H_{10}$, $C_{40}H_{16}$, $C_{76}H_{22}$, $C_{124}H_{28}$, ... (cf. Table 3) are generated. It is only necessary to count a set of symmetrically non-equivalent fissures (marked by arrows below) in the circular benzenoids in question, k-circumphenalenes (scheme on the next page).

Filling of the fissures, one at a time, in the longer rows of hexagons (e.g. at the bottom of the below diagram) creates normal ($\Delta = 0$) benzenoids (**n**). For $k = 0, 1, 2, 3, 4, 5, ...$ the numbers of nonisomorphic such systems are 1, 1, 2, 2, 3, 3, ...; in general: $\lceil (k + 1)/2 \rceil = (1/2)(k + 1) + (1/2)[1 + (-1)^k]$. Similarly for the $\Delta = 1$ systems (o_1), created by filling the fissures in the shorter rows (e.g. at the top of the diagram), one obtains the numbers 0, 1, 1, 2, 2, 3, ..., in general: $\lceil k/2 \rceil = (1/2)(k + 1) - (1/2)[1 + (-1)^k]$. For $k = 0$ the created single **n** system

S. J. Cyvin, B. N. Cyvin, and J. Brunvoll

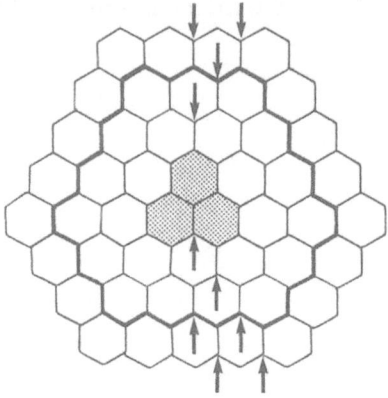

has the symmetry D_{2h}. In all other cases $(k > 0)$ there will be one C_{2v} system, $\mathbf{o}_1$ when k is odd or $\mathbf{n}$ when k is even. All the other systems are unsymmetrical (C_s). The total number of nonisomorphic systems is found among the below equations

$$P(0, 6, k) = 3k + 4 \tag{34}$$

$$P(0, 1, k) = k + \lceil (k + 1)/2 \rceil = (1/2)(3k + 1) + (1/2)[1 + (-1)^k] \tag{35}$$

$$P(0, 2, k) = k + 1 \tag{36}$$

$$P(0, 3, k) = k + 1 + \lceil k/2 \rceil = (1/2)(3k + 3) - (1/2)[1 + (-1)^k] \tag{37}$$

$$P(0, 4, k) = 3k + 3 \tag{38}$$

$$P(-1, 5, k) = \lceil k/2 \rceil = (1/2)(k + 1) - (1/2)[1 + (-1)^k] \tag{39}$$

Table 7 gives a more detailed listing of the P functions, taking into account the classifications according to the Δ values and symmetry.

Table 7. Formulas* for $P(j, \delta, k)$; $k > 0$

j	δ	Δ	C_{2v}	C_s	Total
0	6	0	ϵ	$(1/2)(3k+5-\epsilon)$	$(1/2)(3k+5+\epsilon)$
		2	$1-\epsilon$	$(1/2)(3k+1+\epsilon)$	$(1/2)(3k+3-\epsilon)$
0	1	1	ϵ	$(1/2)(3k+1-\epsilon)$	$(1/2)(3k+1+\epsilon)$
0	2	0	ϵ	$(1/2)(k+1-\epsilon)$	$(1/2)(k+1+\epsilon)$
		2	$1-\epsilon$	$(1/2)(k-1+\epsilon)$	$(1/2)(k+1-\epsilon)$
0	3	1	$1-\epsilon$	$(1/2)(3k+1+\epsilon)$	$(1/2)(3k+3-\epsilon)$
0	4	0	$1-\epsilon$	$(1/2)(3k+3+\epsilon)$	$(1/2)(3k+5-\epsilon)$
		2	ϵ	$(1/2)(3k+1-\epsilon)$	$(1/2)(3k+1+\epsilon)$
-1	5	1	$1-\epsilon$	$(1/2)(k-1+\epsilon)$	$(1/2)(k+1-\epsilon)$

* $\epsilon = (1/2)[1 + (-1)^k]$

12.3 Second Part of the Cardinality

If C_nH_s has an *excised internal structure* [8, 32–34, 38], then its formula is $C_{n-2s+6}H_{s-6}$ in accord with Eq. (26). This is also the formula which appears in point (c′) of the special *aufbau* procedure (Sect. 11). In general an excised internal structure may be a non-benzenoid. Several examples of this phenomenon are found in Fig. 10. In connection with Eq. (27), however, when the excised internal structures of not too small ground forms (G) of constant-isomer benzenoid series are going to be considered, we shall only encounter benzenoids as such systems. Therefore, all we need to known about the definition of an excised internal structure here, is: if $G \equiv$ circum-G^0, then the benzenoid G^0 is the excised internal structure of G. Let again $(n; s)$ be the formula of G, and correspondingly $(n^0; s^0)$ of G^0. Then

$$G^0: \qquad (n^0; s^0) = (n - 2s + 6; s - 6) \tag{40}$$

In Table 8 the $(n^0; s^0)$ formulas are listed as corresponding to the G formulas of Table 3. In spite of the labelling by j and δ also in Table 8, the code $\{j, \delta\}$ applies to G (Table 3), not to G^0 (Table 8). The formulas in parentheses in Table 8, associated with some of the smallest G benzenoids, are not compatible with benzenoids themselves. Firstly, they correspond to the G benzenoids with the six formulas for which the scheme (a) is sufficient for the *aufbau*; they are specified in Sect. 12.1. Their excised internal structures are non-benzenoids. It is interesting to notice that they are nevertheless correctly reproduced by the parenthesized formulas. This is true for, e.g. C_4H_6 butadiene (cf. Fig. 11), C_3H_5 propenyl (Fig. 10), C_2H_4 ethene (Fig. 10), and even for methenyl CH_3, which is represented by a single internal vertex (in $C_{13}H_9$; see Fig. 10). Secondly, it seems reasonable to characterize the two remaining parenthesized formulas of Table 8 (C_0H_2 and C_0H_0) as fictitious.

Where in the periodic table is the formula for G^0 located in relation to G? It should be recalled that $(n; s) \equiv \{j, \delta\}$ pertains to G. Then, on combining (40) with (22) one obtains

$$G^0: \qquad (n^0; s^0) = (6j^2 + (2j + 1)\delta - 1 + 2\lfloor \delta/6 \rfloor; 6j + \delta + 1) \tag{41}$$

Table 8. Formulas for excised internal structures of ground forms of constant-isomer benzenoid series*

δ	j						
	0	1	2	3	4	5	6
1	(C_0H_2)	(C_8H_8)	$C_{28}H_{14}$	$C_{60}H_{20}$	$C_{104}H_{26}$	$C_{160}H_{32}$	$C_{228}H_{38}$
2	(CH_3)	$(C_{11}H_9)$	$C_{33}H_{15}$	$C_{67}H_{21}$	$C_{113}H_{27}$	$C_{171}H_{33}$	$C_{241}H_{39}$
3	(C_2H_4)	$C_{14}H_{10}$	$C_{38}H_{16}$	$C_{74}H_{22}$	$C_{122}H_{28}$	$C_{182}H_{34}$	$C_{254}H_{40}$
4	(C_3H_5)	$C_{17}H_{11}$	$C_{43}H_{17}$	$C_{81}H_{23}$	$C_{131}H_{29}$	$C_{193}H_{35}$	$C_{267}H_{41}$
5	(C_4H_6)	$C_{20}H_{12}$	$C_{48}H_{18}$	$C_{88}H_{24}$	$C_{140}H_{30}$	$C_{204}H_{36}$	$C_{280}H_{42}$
6	(C_7H_7)	$C_{25}H_{13}$	$C_{55}H_{19}$	$C_{97}H_{25}$	$C_{151}H_{31}$	$C_{217}H_{37}$	$C_{295}H_{43}$

* Non-benzenoids and fictitious systems in parentheses; for C_6H_6 $\{-1, 5\}$: (C_0H_0)

S. J. Cyvin, B. N. Cyvin, and J. Brunvoll

We shall compare this formula with the formula for the ground forms with the same δ as in G, but one unit less in j. One finds:

$$\{j - 1, \delta\} \equiv (6(j - 1)^2 + 12(j - 1) + (2j + 1)\delta + 7 + 2\lfloor \delta/6 \rfloor;$$
$$6(j - 1) + \delta + 7)$$
$$= (6j^2 + (2j + 1)\delta + 1 + 2\lfloor \delta/6 \rfloor; 6j + \delta + 1) \qquad (42)$$

Now, on comparing (42) with (41) the simple connection is

$$\{j - 1, \delta\} \equiv (n^0 + 2; s^0) \qquad (43)$$

The answer to the above question is therefore: Look up the formula for the ground forms with the same δ as in G, but one unit less in j. Subtract C_2, which means moving two places up and one place to the right in the periodic table for benzenoids.

Another interpretation of the location of the G^0 formula relates it to the circular benzenoid formula just above the one for $\{j - 1, \delta\}$. If O is the circular benzenoid with the formula given by (29), then its excised internal structure has the formula

$$O^0: \qquad (n - 2s + 5; s - 7) \equiv \{o, \delta', j' - 1\} \qquad (44)$$

It is immediately seen, on comparing with Eq. (40), that the formula (44) is obtained by subtracting CH from the one of G^0. Hence the following alternative answer to our question. Subtract C_3H from the formula of G to arrive at O; find the formula for the excised internal structure of O, viz. O^0; move one place up and one to the right in the periodic table to arrive at G^0.

The two alternative descriptions above are illustrated schematically in the below diagram.

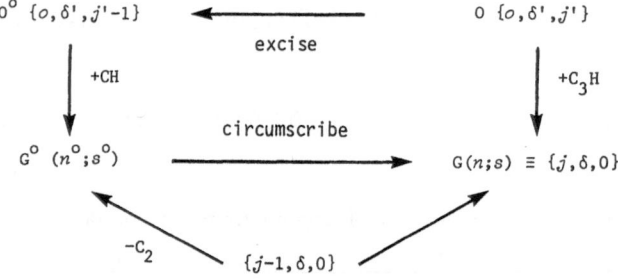

From the above descriptions it should be clear that the G^0 benzenoids, viz. excised internal structures of the ground forms for constant-isomer series, are selected extreme-left benzenoids which are not extremal; there are only some exceptions at the beginning of the periodic table: $C_{14}H_{10}$, $C_{17}H_{11}$, $C_{20}H_{12}$ and $C_{28}H_{14}$ do not belong to extreme-left benzenoids. The formulas for the G^0 benzenoids are marked by stroke symbols in Fig. 9: a cross for the even-carbon atom formulas and the trigonal symbol for the odd-carbon atom formulas.

In accordance with *Proposition 4* (Sect. 6.2) the G^0 benzenoids have no cove and no fjord, at least those which are the selected extreme-left benzenoids. They are found, namely, in rows just above three-formula steps on the staircase-like boundary of the periodic table. But the absence of coves and fjords applies also to the exceptions represented by the four formulas at the beginning of the periodic table as specified above. That may be proved in the same way as *Proposition 4*, but is also found simply by inspection. On the basis of these properties it is inferred that all benzenoids with the formulas $(n^0; s^0)$, which pertain to G^0, can be circumscribed. It seems to be a plausible assumption, although not yet rigorously proved. This discussion is very much the same as the one at the start Sect. 9.1. It is also assumed that the circumscribing of all the $(n^0; s^0)$ benzenoids generate exactly all the $(n; s)$ ground forms which belong to the second part of the cardinality of C_nH_s according to Eq. (27). In other words, it is not necessary to take into account any non-benzenoids.

The final result of this section may be summarized in the equation:

$$|C_nH_s(c')| = |C_{n-2s+6}H_{s-6}| \tag{45}$$

12.4 Conclusion

The cardinality of C_nH_s, which represents the $\{j, \delta\}$ ground forms ($h \geq 13$) for constant-isomer series of benzenoids, is given by

$$|n; s| = P(0 - \lfloor \delta/6 \rfloor, \delta - 1 + 6\lfloor (7 - \delta)/6 \rfloor, j + \lfloor \delta/6 \rfloor$$
$$- \lfloor (7 - \delta)/6 \rfloor) + |n - 2s + 6; s - 6| \tag{46}$$

13 Numerical Results with Finer Classifications

The isomers with the formulas in Table 8 were enumerated by computer aid as far as possible. These systems are excised internal structures (G^0) of ground forms for constant-isomer series of benzenoids when they are benzenoids themselves. They were classified according to Δ values and the symmetry groups to which they belong. The results are listed in Tables 9 and 10 for the even and odd carbon atom formulas, respectively. The documentation (given by footnotes) therein follow the previous review [8].

The numbers for the constant-isomer series of benzenoids, classified according to Δ values and symmetry, are of especially great interest. The symmetry groups for members of constant-isomer series were considered previously at least by Dias [50, 53] in very recent works.

Classified numbers for constant-isomer series pertaining to even-carbon atom formulas are displayed in Table 11. These data (for $h \geq 13$) are accessible by means of Table 9 in combination with the formulas of Table 7, in accordance with the splitting of the cardinalities (27). Correspondingly, the data in Table 12

S. J. Cyvin, B. N. Cyvin, and J. Brunvoll

Table 9. Numbers of isomers, classified according to Δ and symmetry, for even-carbon atom formulas from Table 8*

h	n_i	Formula	Δ	D_{6h}	D_{3h}	C_{3h}	D_{2h}	C_{2h}	C_{2v}	C_s	Total
3	0	$C_{14}H_{10}$	0	0	0	0	1	0	1	0	2[a]
5	2	$C_{20}H_{12}$	0	0	0	0	1	0	1	1	3[a]
8	6	$C_{28}H_{14}$	0	0	0	0	1	2	3	2	8[b]
			2	0	0	0	0	0	0	1	1[c]
12	12	$C_{38}H_{16}$	0	0	0	0	1	2	2	5	10[b]
			2	0	0	0	0	0	1	2	3[b]
16	18	$C_{48}H_{18}$	0	0	1	0	2	3	2	14	22[b]
			2	0	0	0	0	0	2	6	8[b]
21	26	$C_{60}H_{20}$	0	0	0	0	1	1	4	26	32[b]
			2	0	0	0	0	0	3	12	15[b]
27	36	$C_{74}H_{22}$	0	0	0	0	1	1	5	29	36[d]
			2	0	0	0	0	0	2	15	17[d]
33	46	$C_{88}H_{24}$	0	0	0	0	1	2	10	66	79[d]
			2	0	0	1	0	0	2	38	41[d]
			4	0	1	0	0	0	0	0	1[d]
40	58	$C_{104}H_{26}$	0	0	0	0	1	8	12	100	121[d]
			2	0	0	0	0	0	2	62	64[d]
			4	0	0	0	0	0	1	0	1[d]
48	72	$C_{122}H_{28}$	0	0	0	0	1	8	11	105	125[d]
			2	0	0	0	0	0	3	65	68[d]
			4	0	0	0	0	0	1	0	1[d]
56	86	$C_{140}H_{30}$	0	0	0	0	2	11	13	230	256[d]
			2	0	0	0	0	0	8	147	155[d]
			4	0	0	0	0	0	0	2	2[d]
65	102	$C_{160}H_{32}$	0	0	0	0	2	4	17	341	364[d]
			2	0	0	0	0	0	9	223	232[d]
			4	0	0	0	0	0	1	3	4[d]
75	120	$C_{182}H_{34}$	0	0	0	0	2	4	18	346	370[d]
			2	0	0	0	0	0	8	228	236[d]
			4	0	0	0	0	0	1	3	4[d]
85	138	$C_{204}H_{36}$	0	1	0	1	1	6	28	702	739[d]
			2	0	0	0	0	0	6	490	496[d]
			4	0	0	0	0	0	3	9	12[d]

* Symbol —|— in Fig. 9. [a] Elk (1980) [52]; [b] Brunvoll and Cyvin (1990) [20]; [c] Brunvoll, Cyvin, Cyvin and Gutman (1988) [56]; [d] Brunvoll, Cyvin and Cyvin (1991) [8].

(for $h \geq 15$) pertaining to odd-carbon atom formulas are available from Table 10 and the algebraic formulas (Table 7).

In the documentations of Tables 11 and 12 again the previous review [8] was followed; in the cited references the Δ values are specified explicitly. However, several of the distributions into symmetry groups may be extracted from papers by Dias [50, 53] if one identifies non-radicals with $\Delta = 0$ systems, monoradicals with $\Delta = 1$ and diradicals with $\Delta = 2$. This correlation holds in the present cases, but not in general. Thus Dias [50, 53] has specified (among other topological properties) the symmetry groups for the $C_{22}H_{12}$ ($\Delta = 0, 2$) isomers, $C_{30}H_{14}$ ($\Delta = 0, 2$),

Table 10. Numbers of isomers, classified according to Δ and symmetry, for odd-carbon atom formulas from Table 8*

h	n_i	Formula	Δ	D_{3h}	C_{3h}	C_{2v}	C_s	Total
4	1	$C_{17}H_{11}$	1	0	0	0	1	1[a]
7	5	$C_{25}H_{13}$	1	0	0	1	2	3[b]
10	9	$C_{33}H_{15}$	1	0	0	2	6	8[b]
			3	1	0	0	0	1[c]
14	15	$C_{43}H_{17}$	1	0	0	3	12	15[b]
			3	0	0	1	0	1[b]
19	23	$C_{55}H_{19}$	1	0	0	2	15	17[b]
			3	0	0	1	0	1[b]
24	31	$C_{67}H_{21}$	1	0	1	2	38	41[d]
			3	0	0	0	2	2[d]
30	41	$C_{81}H_{23}$	1	0	0	2	62	64[d]
			3	0	0	1	3	4[d]
37	53	$C_{97}H_{25}$	1	0	0	3	65	68[d]
			3	0	0	1	3	4[d]
44	65	$C_{113}H_{27}$	1	0	0	8	147	155[d]
			3	0	0	3	9	12[d]
52	79	$C_{131}H_{29}$	1	0	0	9	223	232[d]
			3	0	0	3	16	19[d]
61	95	$C_{151}H_{31}$	1	0	0	8	228	236[d]
			3	0	0	3	16	19[d]
70	111	$C_{171}H_{33}$	1	0	0	6	490	496[d]
			3	0	1	2	43	46[d]
80	129	$C_{193}H_{35}$	1	0	0	9	708	717[d]
			3	0	0	3	67	70[d]
91	149	$C_{217}H_{37}$	1	0	0	10	713	723[d]
			3	0	0	3	67	70[d]

* Symbol λ in Fig. 9; [a–d] See footnotes to Table 9.

$C_{40}H_{16}$ ($\Delta = 0, 2$), $C_{50}H_{18}$ ($\Delta = 2$), $C_{62}H_{20}$ ($\Delta = 0, 2$), $C_{76}H_{22}$ ($\Delta = 0, 2$), $C_{90}H_{24}$ ($\Delta = 2$), $C_{106}H_{26}$ ($\Delta - 0, 2$), and the $C_{124}H_{28}$ ($\Delta = 0, 2$) isomers; cf. Table 11. In the case of $C_{106}H_{26}/C_{124}H_{28}$ Dias [53] misassigned the D_{2h} system as C_{2v}. Furthermore, Dias [50, 53] specified the symmetry groups for $C_{35}H_{15}$, $C_{45}H_{17}$, $C_{57}H_{19}$, $C_{69}H_{21}$, $C_{83}H_{23}$ and $C_{99}H_{25}$. Finally, Dias [53] identified the following isomers (without indicating the symmetry): the tetraradicals ($\Delta = 4$), one each of $C_{142}H_{30}$, $C_{162}H_{32}$ and $C_{184}H_{34}$ (cf. Table 11); two triradicals ($\Delta = 3$) $C_{115}H_{27}$ and four each of $C_{133}H_{29}$ and $C_{153}H_{31}$ (cf. Table 12).

In Tables 10 and 12 there are no columns for the symmetries D_{6h}, C_{6h}, D_{2h} and C_{2h}. The absence of benzenoids with these symmetries for odd-carbon atom formulas is consistent with the fact that all such systems have $\Delta = 0$. The reason why the column for C_{6h} is missing in Tables 9 and 11 is explained in the next section.

S. J. Cyvin, B. N. Cyvin, and J. Brunvoll

Table 11. Numbers of isomers, classified according to Δ and symmetry, for constant-isomer benzenoid series with even-carbon atom formulas

h	n_i	Formula*	Δ	D_{6h}	D_{3h}	C_{3h}	D_{2h}	C_{2h}	C_{2v}	C_s	Total
1	0	C_6H_6	0	1	0	0	0	0	0	0	1[a]
2	0	$C_{10}H_8$	0	0	0	0	1	0	0	0	1[a]
4	2	$C_{16}H_{10}$	0	0	0	0	1	0	0	0	1[a]
6	4	$C_{22}H_{12}$	0	0	0	0	0	1	1	0	2[a]
			2	0	1	0	0	0	0	0	1[b]
9	8	$C_{30}H_{14}$	0	0	0	0	1	0	1	1	3[a]
			2	0	0	0	0	0	1	0	1[a]
13	14	$C_{40}H_{16}$	0	0	0	0	1	0	1	1	3[a]
			2	0	0	0	0	0	1	0	1[a]
17	20	$C_{50}H_{18}$	0	0	0	0	1	0	2	4	7[a]
			2	0	0	0	0	0	0	2	2[a]
22	28	$C_{62}H_{20}$	0	0	0	0	1	2	3	6	12[a]
			2	0	0	0	0	0	1	3	4[a]
28	38	$C_{76}H_{22}$	0	0	0	0	1	2	3	6	12[a]
			2	0	0	0	0	0	1	3	4[a]
34	48	$C_{90}H_{24}$	0	0	1	0	2	3	2	19	27[a]
			2	0	0	0	0	0	3	9	12[a]
41	60	$C_{106}H_{26}$	0	0	0	0	1	1	5	31	38[a]
			2	0	0	0	0	0	3	16	19[a]
49	74	$C_{124}H_{28}$	0	0	0	0	1	1	5	31	38[c]
			2	0	0	0	0	0	3	16	19[c]
57	88	$C_{142}H_{30}$	0	0	0	0	1	2	11	72	86[c]
			2	0	0	1	0	0	2	43	46[c]
			4	0	1	0	0	0	0	0	1[c]
66	104	$C_{162}H_{32}$	0	0	0	0	1	8	12	107	128[c]
			2	0	0	0	0	0	3	67	70[c]
			4	0	0	0	0	0	1	0	1[c]
76	122	$C_{184}H_{34}$	0	0	0	0	1	8	12	107	128[c]
			2	0	0	0	0	0	3	67	70[c]
			4	0	0	0	0	0	1	0	1[c]
86	140	$C_{206}H_{36}$	0	0	0	0	2	11	13	238	264[c]
			2	0	0	0	0	0	9	153	162[c]
			4	0	0	0	0	0	0	2	2[c]
97	160	$C_{230}H_{38}$	0	0	0	0	2	4	18	349	373
			2	0	0	0	0	0	9	230	239
			4	0	0	0	0	0	1	3	4
109	182	$C_{256}H_{40}$	0	0	0	0	2	4	18	349	373
			2	0	0	0	0	0	9	230	239
			4	0	0	0	0	0	1	3	4
121	204	$C_{282}H_{42}$	0	1	0	1	1	6	29	711	749
			2	0	0	0	0	0	6	498	504
			4	0	0	0	0	0	3	9	12

Table 12. Numbers of isomers, classified according to Δ and symmetry, for constant-isomer benzenoid series with odd-carbon atom formulas

Ground form								
h	n_i	Formula*	Δ	D_{3h}	C_{3h}	C_{2v}	C_s	Total
3	1	$C_{13}H_9$	1	1	0	0	0	1[a]
5	3	$C_{19}H_{11}$	1	0	0	1	0	1[b]
8	7	$C_{27}H_{13}$	1	0	0	1	0	1[b]
11	11	$C_{35}H_{15}$	1	0	0	0	2	2[b]
15	17	$C_{45}H_{17}$	1	0	0	1	3	4[b]
20	25	$C_{57}H_{19}$	1	0	0	1	3	4[b]
25	33	$C_{69}H_{21}$	1	0	0	3	9	12[b]
			3	1	0	0	0	1[b]
31	43	$C_{83}H_{23}$	1	0	0	3	16	19[b]
			3	0	0	1	0	1[b]
38	55	$C_{99}H_{25}$	1	0	0	3	16	19[b]
			3	0	0	1	0	1[b]
45	67	$C_{115}H_{27}$	1	0	1	2	43	46[c]
			3	0	0	0	2	2[c]
53	81	$C_{133}H_{29}$	1	0	0	3	67	70[c]
			3	0	0	1	3	4[c]
62	97	$C_{153}H_{31}$	1	0	0	3	67	70[c]
			3	0	0	1	3	4[c]
71	113	$C_{173}H_{33}$	1	0	0	9	153	162[c]
			3	0	0	3	9	12[c]
81	131	$C_{195}H_{35}$	1	0	0	9	230	239[c]
			3	0	0	3	16	19[c]
92	151	$C_{219}H_{37}$	1	0	0	9	230	239[c]
			3	0	0	3	16	19[c]
103	171	$C_{243}H_{39}$	1	0	0	6	498	504
			3	0	1	2	43	46
115	193	$C_{269}H_{41}$	1	0	0	10	716	726
			3	0	0	3	67	70
128	217	$C_{297}H_{43}$	1	0	0	10	716	726
			3	0	0	3	67	70

* Symbol ▲ in Fig. 9; [a] Brunvoll, Cyvin, Cyvin and Gutman (1988) [56]; [b] Brunvoll and Cyvin (1990) [20]; [c] Brunvoll, Cyvin and Cyvin (1991) [8].

◄ (Note to Table 11)

* Symbol ◆ in Fig. 9; [a] Brunvoll and Cyvin (1990) [20]; [b] Brunvoll, Cyvin, Cyvin and Gutman (1988) [56]; [c] Brunvoll, Cyvin and Cyvin (1991) [8].

14 Constant-Isomer Benzenoid Series: Snowflakes

A *snowflake* in the present sense is a benzenoid of hexagonal symmetry [8, 57, 58], belonging to D_{6h} or C_{6h}. Sometimes the symmetries D_{6h} and C_{6h}, are associated to *proper-* and *improper snowflakes* [59], respectively.

From Table 11 it is seen that the snowflakes are sparsely distributed among the constant-isomer series of benzenoids. Apart from benzene ($h = 1$) there is one single ground form of D_{6h} symmetry among the tabulated systems. This next-smallest ground form snowflake is $C_{282}H_{42}$ ($h = 121$) tetracircum-hexabenzo[bc, ef, hi, kl, no, qr]coronene; see Fig. 16. It was found that the smallest C_{6h} ground

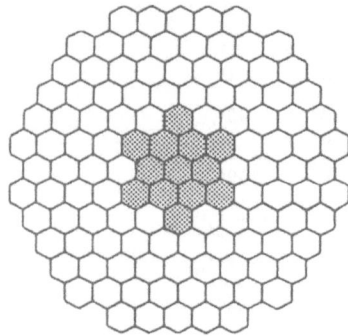

Fig. 16. The next-smallest ground form of constant-isomer benzenoid series (after benzene C_6H_6) which is a proper snowflake: $C_{282}H_{42}$ ($h = 121$)

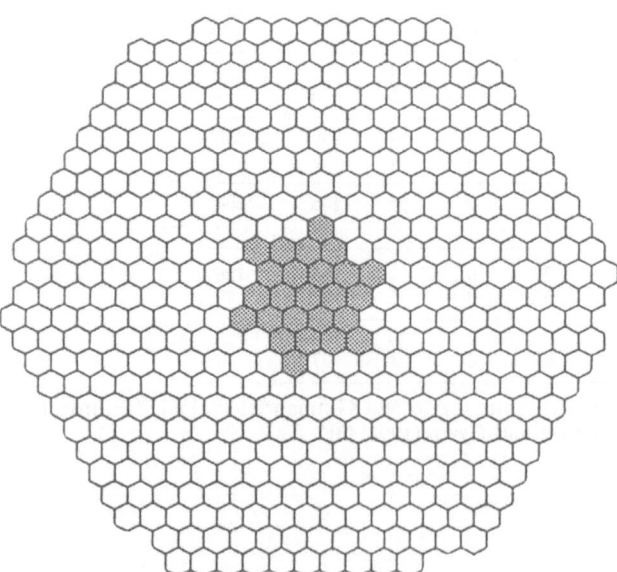

Fig. 17. The smallest ground form of constant-isomer benzenoid series which is an improper snowflake: $C_{990}H_{78}$ ($h = 457$)

Table 13. Numbers of snowflakes among constant-isomer benzenoid series

Ground form		Formula	D_{6h}	C_{6h}
h	n_i			
1	0	C_6H_6	1	0
121	204	$C_{282}H_{42}$	1	0
457	840	$C_{990}H_{78}$	0	1
1009	1908	$C_{2130}H_{114}$	1	1

form occurs for $h = 457$. It is a nonacircum-hexabenzocircumcoronene as shown in Fig. 17 and far beyond the listing of Table 11. It is therefore understandable that the column for C_{6h} symmetry was skipped therein. Correspondingly, the smallest C_{6h} system in continuation of Table 9 would appear as $C_{840}H_{72}$ at $h = 385$.

Table 13 summarizes and supplements the findings about snowflakes described above.

15 Dias Paradigm and Other Regularities Among Constant-Isomer Benzenoid Series

In Tables 4 and 5 a spectacular pattern in the numbers is exhibited: in each of the tables one finds the progression of singlets and doublets as a, b, b, c, d, d, e, f, f, ... The pattern applies also to the Kekuléan and non-Kekuléan systems separately (cf. Table 4). It was Dias [50, 53] who first observed these regularities · and referred to them as a "new topological paradigm" [50]. In addition, he described [50, 53] topological equivalences between benzenoids governed by this *Dias paradigm*. A part of these equivalences imply the symmetry groups and the radical/non-radical nature. The latter property is here expressed in terms of Δ values. From Tables 11 and 12 it is seen that the Dias paradigm is valid for the Δ values separately and also for the symmetry distributions within each Δ. For example, the entries for $C_{30}H_{14}/C_{40}H_{16}$ (see Table 11) are entirely identical; the same is the case for $C_{83}H_{23}/C_{99}H_{25}$ (Table 12).

Let a constant-isomer series of benzenoids be identified by the symbol $\{j, \delta\}$; cf. Sect. 9.4 and Table 3. Strictly speaking this symbol was specifically used for the ground forms. However, the discussion of the present section is not restricted to these systems since the properties under consideration (Δ and symmetry) do not change by circumscribing. Let the partial cardinality for $\{j, \delta\}$ at a given Δ value be written $|\{j, \delta\}, \Delta|$. Then the Dias paradigm in a strict sense, as will be used presently, is expressed by:

$$|\{j, 1\}, \Delta| = |\{j, 3\}, \Delta|; \quad \Delta = 0, 2, 4, 6, \ldots \tag{47}$$

$$|\{j, 4\}, \Delta| = |\{j, 6\}, \Delta|; \quad \Delta = 1, 3, 5, 7, \ldots \tag{48}$$

Table 14. Illustration of the equivalence of partial cardinalities for $\Delta = 1, 2, 3, 4$

Δ; (s shift in parentheses)				
1; (0)	2; (3)	3; (12)	4; (21)	Number
$\lvert C_{13}H_9 \rvert$	$\lvert C_{22}H_{12} \rvert$	$\lvert C_{69}H_{21} \rvert$	$\lvert C_{142}H_{30} \rvert$	1
$\lvert C_{19}H_{11} \rvert$	$\lvert C_{30}H_{14} \rvert$	$\lvert C_{83}H_{23} \rvert$	$\lvert C_{162}H_{32} \rvert$	1
$\lvert C_{27}H_{13} \rvert$	$\lvert C_{40}H_{16} \rvert$	$\lvert C_{99}H_{25} \rvert$	$\lvert C_{184}H_{34} \rvert$	1
$\lvert C_{35}H_{15} \rvert$	$\lvert C_{50}H_{18} \rvert$	$\lvert C_{115}H_{27} \rvert$	$\lvert C_{206}H_{36} \rvert$	2
$\lvert C_{45}H_{17} \rvert$	$\lvert C_{62}H_{20} \rvert$	$\lvert C_{133}H_{29} \rvert$	$\lvert C_{230}H_{38} \rvert$	4
$\lvert C_{57}H_{19} \rvert$	$\lvert C_{76}H_{22} \rvert$	$\lvert C_{153}H_{31} \rvert$	$\lvert C_{256}H_{40} \rvert$	4
$\lvert C_{69}H_{21} \rvert$	$\lvert C_{90}H_{24} \rvert$	$\lvert C_{173}H_{33} \rvert$	$\lvert C_{282}H_{42} \rvert$	12
$\lvert C_{83}H_{23} \rvert$	$\lvert C_{106}H_{26} \rvert$	$\lvert C_{195}H_{35} \rvert$	$\lvert C_{310}H_{44} \rvert$	19
$\lvert C_{99}H_{25} \rvert$	$\lvert C_{124}H_{28} \rvert$	$\lvert C_{219}H_{37} \rvert$	$\lvert C_{340}H_{46} \rvert$	19

The equivalence in symmetry distributions is tacitly assumed to be valid in relations of this type.

Dias [53] has also observed some interesting regularities in the numbers for constant-isomer benzenoid series which combine the even-carbon atom formulas (Table 11) with the odd-carbon atom formulas (Table 12). Here we express these regularities by:

$$\lvert \{j, 2\}, \Delta = 1 \rvert = \lvert \{j, 5\}, \Delta = 2 \rvert \tag{49}$$

$$\lvert \{j, 4\}, \Delta = 1 \rvert = \lvert \{j + 1, 1\}, \Delta = 2 \rvert$$
$$= \lvert \{j, 6\}, \Delta = 1 \rvert = \lvert \{j + 1, 3\}, \Delta = 2 \rvert \tag{50}$$

Below we specify additional regularities, which also involve $\Delta = 3$ and $\Delta = 4$. For the odd-carbon atom formulas it was observed

$$\lvert \{j, \delta\}, \Delta = 1 \rvert = \lvert \{j + 2, \delta\}, \Delta = 3 \rvert ; \qquad \delta = 2, 4, 6 \tag{51}$$

and for the even-carbon atom formulas:

$$\lvert \{j, \delta\}, \Delta = 2 \rvert = \lvert \{j + 3, \delta\}, \Delta = 4 \rvert ; \qquad \delta = 1, 3, 5 \tag{52}$$

As a consequence of the relations (50)–(52) a certain sequence of numbers repeats itself for $\Delta = 1, 2, 3$ and 4, only with a constant shift in the s value. Thus, for instance, the partial cardinalities $\lvert C_{13}H_9 \rvert$, $\lvert C_{19}H_{11} \rvert$, $\lvert C_{27}H_{13} \rvert$, $\lvert C_{35}H_{15} \rvert$, ... with $\Delta = 1$ (actually complete for $h < 20$) are pairwise equal to the partial cardinalities for $\lvert C_{22}H_{12} \rvert$, $\lvert C_{30}H_{14} \rvert$, $\lvert C_{40}H_{16} \rvert$, $\lvert C_{50}H_{18} \rvert$, ... with $\Delta = 2$. Here the shift in s is 3 units. Table 14 shows a survey of these properties. All the data in this table are extracted from Tables 11 and 12 except the two last numbers, which are extrapolations, viz. $|C_{310}H_{44}| = |C_{340}H_{46}| = 19$ at $\Delta = 4$.

16 Conclusion

The present chapter should be regarded as a continuation of a previous chapter [8] in the same monograph series. These two chapters together provide an extensive review of most of the enumerations of benzenoid isomers, characterized by the (chemical) formulas C_nH_s. A specific treatment of all-benzenoids [13, 60, 61] is not included here, although several treatments of their numbers of isomers (explicitly as C_nH_s) are available [23, 34, 36, 38, 40, 41, 55].

Emphasis is laid on the constant-isomer series of benzenoids, not at least on the relevant enumerations in this connection. This is a typical branch of computational chemistry. A substantial amount of new computational results are reported, as well as original contributions to the background theory in the area.

There is still more to be done in the studies of benzenoid isomers and the constant-isomer series in particular. The Dias paradigm is interesting, as well as the accompanying topological characteristics of the pertinent benzenoids. It is stressed that these patterns have not been proved rigorously. The new regularities presented in Sect. 15 have not been proved, but this might be attempted. Furthermore, it would be interesting to extend these regularities, if possible, to higher Δ values, viz. $\Delta > 4$. Thus there are still many problems to be solved in connection with the constant-isomer series of benzenoids, but intensive research in this area is in progress. References are given to very recent relevant papers [62–69], which are not cited elsewhere in this chapter.

For a final comment we go back to the alkane C_NH_{2N+2} isomers mentioned in the Introduction. Davies and Freyd [70] have reported as the largest number of such isomers $\sim 1.35 \times 10^{173}$ (given to 15 significant figures) for $C_{400}H_{802}$. The number $\sim 1.6 \times 10^{80}$ for $N = 167$ is characterized in the title of the paper [70]: $C_{167}H_{336}$ Is the Smallest Alkane with More Realizable Isomers than the Observed Universe Has "Particles".

Acknowledgement. Financial support to BNC from The Norwegian Research Council for Science and the Humanities is gratefully acknowledged.

17 References

1. Cayley A (1875) Ber Dtsch Chem Ges 8: 172
2. Herrmann F (1880) Ber Dtsch Chem Ges 13: 792
3. Trinajstić N (1983) Chemical graph theory, vol I, II, CRC Press, Boca Raton, Florida
4. Hence HR, Blair CM (1931) J Am Chem Soc 53: 3077
5. Davis CC, Cross K, Ebel M (1971) J Chem Educ 48: 675
6. Knop JV, Müller WR, Jeričević Ž, Trinajstić N (1981) J Chem Inf Comput Sci 21: 91

7. Knop JV, Müller WR, Szymanski K, Trinajstić N (1985) Computer generation of certain classes of molecules, SKTH/Kemija u industriji (Association of Chemists and Technologists of Croatia), Zagreb
8. Brunvoll J, Cyvin BN, Cyvin SJ (1992) In: Gutman I (ed) Advances in the theory of benzenoid hydrocarbons II (Topics in Current Chemistry 162), Springer, Berlin Heidelberg New York, p 181
9. Gutman I (1982) Bull Soc Chim Beograd 47: 453
10. Gutman I, Polansky OE (1986) Mathematical concepts in organic chemistry, Springer, Berlin Heidelberg New York
11. Cyvin SJ, Gutman I (1988) Kekulé structures in benzenoid hydrocarbons (Lecture Notes in Chemistry 46), Springer, Berlin Heidelberg New York
12. Gutman I, Cyvin SJ (1989) Introduction to the theory of benzenoid hydrocarbons, Springer, Berlin Heidelberg New York
13. Cyvin BN, Brunvoll J, Cyvin SJ (1992) In: Gutman I (ed) Advances in the theory of benzenoid hydrocarbons II (Topics in Current Chemistry 162), Springer, Berlin Heidelberg New York
14. Polansky OE, Rouvray DH (1976) Match 2: 63
15. Cyvin SJ, Brunvoll J, Cyvin BN (1991) Theory of coronoid hydrocarbons (Lecture Notes in Chemistry 54), Springer, Berlin Heidelberg New York
16. Dias JR (1982) J Chem Inf Comput Sci 22: 15
17. Stojmenović I, Tošić R, Doroslovački R (1986) In: Tošić R, Acketa D, Petrović V (eds) Graph theory, Proceedings of the sixth Yugoslav seminar on graph theory, Dubrovnik, April 18–19, 1985, University of Novi Sad, Novi Sad, p 189
18. Doroslovački R, Tošić R (1984) Reviews of Research Fac Sci Univ Novi Sad, Mathematics Series 14: 201
19. Tošic R, Doroslovački R, Gutman I (1986) Match 19: 219
20. Brunvoll J, Cyvin SJ (1990) Z Naturforsch 45a: 69
21. Knop JV, Szymanski K, Jeričević Ž, Trinajstić N (1983) J Comput Chem 4: 23
22. Trinajstić N, Jeričević Ž, Knop JV, Müller WR, Szymanski K (1983) Pure & Appl Chem 55: 379
23. Cyvin SJ, Brunvoll J, Cyvin BN (1991) J Math Chem 8: 63
24. Brunvoll J, Cyvin SJ, Cyvin BN (1991) J Mol Struct (Theochem) 235: 147
25. Müller WR, Szymanski K, Knop JV, Nikolić S, Trinajstić N (1989) Croat Chem Acta 62: 481
26. Knop JV, Müller WR, Szymanski K, Trinajstić N (1990) 30: 159
27. Müller WR, Szymanski K, Knop JV, Nikolić S, Trinajstić N (1990) J Comput Chem 11: 223
28. Nikolić S, Trinajstić N, Knop JV, Müller WR, Szymanski K (1990) J Math Chem 4: 357
29. Knop JV, Müller WR, Szymanski K, Nikolić S, Trinajstić N (1990) In: Rouvray DH (ed) Computational graph theory, Nova, New York, p 9
30. Cyvin SJ, Brunvoll J (1990) Chem Phys Letters 170: 364
31. Brunvoll J, Cyvin SJ, Cyvin BN (1987) J Comput Chem 8: 189
32. Dias JR (1989) Z Naturforsch 44a: 765
33. Dias JR (1990) J Math Chem 4: 17
34. Dias JR (1990) In: Gutman I, Cyvin SJ (eds) Advances in the theory of benzenoid hydrocarbons (Topics in Current Chemistry 153), Springer, Berlin Heidelberg New York p 123
35. Dias JR (1983) Match 14: 83
36. Dias JR (1985) Accounts Chem Res 18: 241
37. Dias JR (1986) J Mol Struct (Theochem) 137: 9
38. Dias JR (1987) Handbook of polycyclic hydrocarbons – Part A – Benzenoid hydrocarbons, Elsevier, Amsterdam
39. Dias JR (1984) J Chem Inf Comput Sci 24: 124
40. Dias JR (1989) J Mol Struct (Theochem) 185: 57
41. Dias JR (1987) Thermochim. Acta 122: 313
42. Cyvin BN, Brunvoll J, Cyvin SJ and Gutman I (1988) Match 23: 163

43. Cyvin SJ, Brunvoll J, Cyvin BN (1991) Match 26: 27
44. Harary F, Harborth H (1976) J Combinat Inf System Sci 1: 1
45. Dias JR (1984) Can J Chem 62: 2914
46. Dias JR (1990) J Chem Inf Comput Sci 30: 61
47. Brunvoll J, Cyvin BN, Cyvin SJ, Gutman I, Tošić R, Kovačević M (1989) J Mol Struct (Theochem) 184: 165
48. Balaban AT (1971) Tetrahedron 27: 6115
49. Cyvin SJ, Brunvoll J, Gutman I (1990) Rev Roumaine Chim 35: 985
50. Dias JR (1990) J Chem Inf Comput Sci 30: 251
51. Cyvin SJ, Brunvoll J (1991) Chem Phys Letters 176: 413
52. Elk SB (1980) Match 8: 121
53. Dias JR (1990) Theor Chim Acta 77: 143
54. Dias JR (1985) J Macromol Sci, Chem A 22: 335
55. Dias JR (1985) Nouv J Chim 9: 125
56. Brunvoll J, Cyvin BN, Cyvin SJ, Gutman I (1988) Z Naturforsch 43a: 889
57. Hosoya H (1986) Computers Math Applic 12B: 271; reprinted in: Hargittai I (ed) (1986) Symmetry unifying human understanding, Pergamon, New York
58. Cyvin SJ, Bergan JL, Cyvin BN (1987) Acta Chim Hung 124: 691
59. Cyvin SJ, Brunvoll J, Cyvin BN (1989) Computers Math Applic 17: 355; printed as: Hargittai I (ed) (1989) Symmetry 2 unifying human understanding, Pergamon, Oxford
60. Polansky OE, Derflinger G (1967) Internat J Quant Chem 1: 379
61. Polansky OE, Rouvray DH (1976) Match 2: 91
62. Dias JR (1990) Z Naturforsch 45a: 1335
63. Dias JR (1991) Chem Phys Letters 176: 559
64. Cyvin SJ (1991) Chem Phys Letters 181: 431
65. Cyvin SJ (1991) Coll Sci Papers Fac Sci Kragujevac 12: 95
66. Dias JR (1991) J Chem Inf Comput Sci (1991) 31: 89
67. Cyvin SJ (1991) J Chem Inf Comput Sci 31: 340
68. Cyvin SJ, Brunvoll J, Cyvin BN (1991) J Serb Chem Soc 56: 369
69. Cyvin SJ (1991) Theor Chim Acta 81: 269
70. Davies RE, Freyd PJ (1989) J Chem Educ 66: 278

The Synthon Model and the Program PEGAS for Computer Assisted Organic Synthesis

Eva Hladká[1], Jaroslav Koča[1], Milan Kratochvíl[1], Vladimír Kvasnička[2],
Luděk Matyska[3], Jiří Pospíchal[2], and Vladimír Potůček[1]

Table of Contents

[1] Department of Organic Chemistry, Faculty of Science, Masaryk University, 61137 Brno, Czechoslovakia
[2] Department of Mathematics, Faculty of Chemical Technology, Slovak Technical University, 81237 Bratislava, Czechoslovakia
[3] Institute of Computer Science, Masaryk University, 60177 Brno, Czechoslovakia

Topics in Current Chemistry, Vol. 166
© Springer-Verlag Berlin Heidelberg 1993

A synthon model of organic chemistry is described. A synthon is determined as a molecular graph in which a certain part of vertices is distinguished and those vertices are called the virtual vertices. The virtual vertices correspond to functional groups or molecular fragments that do not directly participate in chemical transformations (though these transformations may be substantially influenced by them). The synthon precursors or successors are formed from a given synthon by special sort of reaction graph in such a way that all valence states of atoms (vertices) in a precursor or successor are classified as stable valence states. The program PEGAS represents an implementation of the synthon model. Its basic features, scopes, and algorithms are briefly outlined. The effectiveness of the program PEGAS is illustrated by many examples of interest for organic chemists.

1 Introduction

Mathematical chemistry, the new challenging discipline of chemistry has established itself in recent years. Its main goal is to develop formal (mathematical) methods for chemical theory and (to some extent) for data analysis. Its history may be traced back to Caley's attempt, more than 100 years ago, to use the graph theoretical representation and interpretation of the chemical constitution of molecules for the enumeration of acyclic chemical structures. Graph theory and related areas of discrete mathematics are the main tools of qualitative theoretical treatment of chemistry [1, 2]. However, attempts to contemplate connections between mathematics and chemistry and to predict new chemical facts with the help of formal mathematics have been scarce throughout the entire history of chemistry.

Mathematics was incorporated into chemistry from physics, and for a long time it was used mainly as a tool for calculations. There is no doubt that such an approach led to important discoveries, and numerical computations are of great importance. Quantum chemistry may be seen as a good example of advances in theoretical physics, and numerical methods which together with the development of new computer hardware allowed us to calculate many properties of molecules with great accuracy. However, quantum chemistry is, in essence, the theoretical physics of molecules, and may be classified as the physical "metatheory" of chemistry which interprets phenomenological notions and concepts of chemistry and provides its physical background.

Mathematical chemistry represents a different approach to the development of the formal logical structure of chemistry. It is not meant to change chemistry into physics, but allows rationalization, unification, and formalization of its principal notions and concepts. An analog may be found in, e.g., formal genetics and mathematical linguistics, where finding and expressing connections between the principal notions and concepts by making use of formal tools belongs to the very important present achievements. The topology of chemical knots, the theory of chemical clusters, the study of formal chemical kinetics, the study of theoretical stereochemistry, new uses of topological indices, the mathematical analysis of polymer structures, and last but not least, mathematical models of organic chemistry are currently hot topics in mathematical chemistry.

A very important role in the proliferation and practical use of mathematical chemistry has been played by computers, especially in the last two decades, when the use of computers was extended from numerical computations and data handling to decision-making and logical problem-solving processes. The DENDRAL project, started at Stanford in the 1960s, may serve as an illustrative example [3]. It was fairly successful, and strongly influenced the collaboration between chemists, computer scientists, and mathematicians. Many applications of artificial intelligence in chemistry, and especially the construction of expert systems, are natural consequences of this collaboration [4].

At almost the same time as the DENDRAL project began, Vleduts was the first to put forward the idea of obtaining suggestions for organic synthesis from computers [5], and the famous article of Corey and Wipke [6] started the great adventure of computer aided organic synthesis design (CAOS). Many research groups have entered this field since then (for a review see Ref. [7]).

The solution of CAOS problems undoubtedly requires some kind of intelligence, as the design of organic synthesis of nontrivial chemical compounds is a very complex task, and even the great masters of the field do not know exactly how they achieve their success. Although this may be seen as one of the main driving forces of research, it makes the research very difficult. We have no reasonable possibility of solving the design of organic synthesis by explicit physical computation, and as long as the problem of chemical reactivity is not adequately solved, there will be always room for the creativity (and subjectivity) of chemists (and computer programs, too).

Many of the research groups involved in CAOS are trying to develop not only working programs for forward synthesis and/or retrosynthesis, but also find the underlying concepts and notions, developing thus more or less formal models of organic chemistry. We may classify the research groups and their activities on the basis of how explicitly they express the idea of building the formal model.

On the one side, the database oriented programs may be found, whereas the logically oriented ones are on the other side. These two approaches represent the limiting points in the range of developed computer programs.

The database oriented programs utilize the reaction databases for obtaining the information whether, and how, a given compound reacts. The full information about the compound and its reactions, together with the references to the literature, may be obtained from appropriate databases, but the approach is strictly limited to known reactions of known compounds.

The separation of reactions from the molecules represents the next level of sophistication. Only those parts of molecules, which are changed during the reaction are considered. These parts, called synthons, are stored in the database together with the information about the appropriate reactions. When the synthon is found in a compound under investigation, the reaction (transformation) is always applied, regardless of its plausibility. The information about the environment of the synthon in the reacting compound may also be stored, so the transformation will not be applied to all compounds containing the synthon. This approach cannot lead to a discovery of principally new reactions (as they represent the unmodifiable content of the database used), but the synthesis of new compounds may be suggested, as the reactions are applied to the unusual structures. LHASA [6] and SECS [8] are the best known representatives of this class of CAOS programs. Both programs use specialized languages for representing chemical transformations, and for the rating of changes of reactivity in respect of the changes of the environment of the transformed synthon.

CAMEO may be seen as a next step towards the logically based systems [9]. Although it mimics the reasoning of organic chemists for the forward synthesis, it is based on the belief that few fundamental mechanistic steps and their combinations are sufficient to cover the immense area of organic chemistry. Half-reactions, introduced by Hendrickson [10, 11], allow more abstraction. Any reaction is created from two half-reactions, and the parts of the structures being modified are called half-spans. Although representable in terms of electron moves, the half-reaction representation does not imply any mechanistic considerations. The program MASSO [12], based on this representation, is able to produce many highly innovative propositions (but also many unacceptable ones), as the half-reaction approach overgeneralizes all the reaction schemes already known in synthetic chemistry thus enabling the prediction of not yet explored reactions.

Ugi and his group introduced an algebraic representation of the logical structure of constitutional chemistry in the early 1970s [13]. This global qualitative mathematical model was then used as a formal basis for the development of systems for CAOS in both forward and backward (retro-) syntheses, as there is no distinction between these two directions from the model point of view. The Dugundji-Ugi (DU) model [13] relies on the extension of isomerism from molecules to ensembles of molecules, chemical reactions being formal isomerization between two appropriate ensembles. This model, in principle, is able to represent any reaction allowed by the valence states of atoms. As the isomerization (i.e. the reaction) may be of arbitrary complexity, any reaction mechanism, even not yet described, may be modeled. The DU-model may become the formal framework for the reactions, reaction mechanisms, and their representations for storing and retrieval purposes. The program TOSCA [14], and the program FLAMINGOES [15], developed by the Moscow group headed by Zefirov and Tratch, although not based on the DU-model, also belong to the category of logically oriented programs.

The extension of the DU-model, based on the graph theoretical notions and concepts, was developed by our group [16]. This new model of organic chemistry represented by the so-called synthon model, also serves as a theoretical basis for the implementation of the program PEGAS [17].

This contribution is devoted to the representation of such a new formal mathematical model of organic chemistry. Although the emphasis is given to the use of this model for the implementation of CAOS programs, the model itself represents a background of the logical structure of chemistry, and it is the extension and elaboration of the above mentioned DU-model. One of the main reasons for its suggestion is an attempt to relate the synthon model to the reasoning of organic chemists when they suggest an organic synthesis. A synthon is composed of atoms and the so-called virtual atoms (which represent functional groups not directly participating in chemical reactions), chemical bonds between them, and lone electrons localized on atoms. Each atom of the synthon is in an actual valence state from a list of valence states formed by all chemical elements involved. Mathematical tools have been developed to make it possible to formalize properly the basic concepts of organic syntheses, e.g. the precursor and/or successor of a molecular system. A chemical reaction between two synthons is unambiguously described by the so-called reaction graph, it describes bonds that are annihilated (from an educt synthon) as well as bonds that are created in the course of the reaction studied.

Starting from a given educt synthon, we can form a set of all its possible synthon precursors/successors (SPS) that are generated by simple reaction graphs. It means that an over-all chemical reaction, passing through several SPS can be expressed as a sequence of the above mentioned reaction graphs.

The synthon model is implemented in the form of program PEGAS for personal computers (IBM PC/AT compatible). Its detailed description will be presented in Sect. 4. The results obtained indicate that it offers a very promising computer tool for organic chemists when they are looking for products which could be produced from a given educt molecule, and vice versa, when searching for precursors of a given product.

The model outlined above and its simple extension contain a theoretical possibility of how to direct the production of SPS towards the required product synthon. This very serious problem is partially overcome by making use of the concept of reaction distance; only those new SPS are taken into account that have a smaller distance than their predecessor with respect to the product synthon. Recently, our activity in this field has been concentrated on an application of tools of logical programming for implementation of some principal rules of chemical reactivity.

2 A Simplified Graph-Theoretical Model of Synthons

The formal theory of synthons has been recently elaborated by Koča [18–22] as a generalization of the matrix model of the constitutional chemistry initially introduced by Dugundji and Ugi [13]. In the framework of the synthon approach it is possible to formalize in sketch form the chemical transformations (reactions) that are not a priori stoichiometric. That is, in the course of a chemical transformation, the numbers of atoms and valence electrons are not conserved. The synthon represents that minimal fragment of the molecular system which is necessary for the given chemical transformation. The concept of the synthon for purposes of computer assisted organic synthesis design was initially formulated by Corey [23] in 1967. The most important feature of the synthon approach is that the constraint of strict stoichiometry is removed and the virtual vertices (atoms) are formally treated as functional groups not participating directly in the chemical transformation, though they may substantially affect chemical reactivity.

The topic of the present section is a formulation of an alternative graph-theoretical model of synthons [16, 24]. Such an approach makes it possible to build-up all concepts and notions of synthon theory [18–22] in a very transparent and easily understandable level.

2.1 Synthons

The concept of molecular graph was outlined in our previous communications [16, 25]. This term will be generalized in such a way that some vertices are distinguished from other ones and they are called the *virtual vertices*. These virtual vertices are assembled in a separate *virtual vertex set* $W = \{w, w', ...\}$. The remaining vertices form the *vertex set* (non empty) $A = \{v, v', ...\}$. For simplicity, in the following chapter we shall not consider the description of vertices by atomic labels, but we suppose that they have a defined description. Virtual vertices have

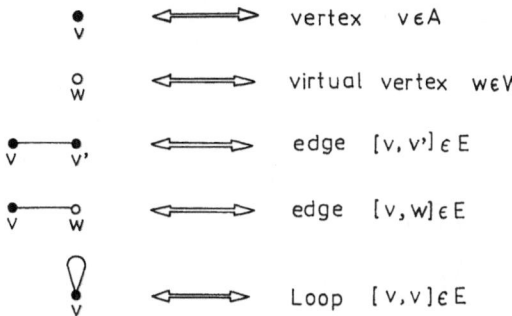

Fig. 1. Diagrammatic representation of single components of the synthon. The vertex from A is represented by a *heavy dot* whereas the virtual vertex from W is represented by an *open circle*. The edge $[v, v']$ ($[v, w]$) is represented by a *continuous line* connecting the vertex v with the vertex $v'(w)$. The loop $[v, v]$ is represented by a *continuous line beginning and ending at the same vertex v*

also no labelling, but in contrast to vertices, virtual vertices can be assigned by arbitrary description, like an unspecified group X in a reaction.

The *edge set* $E = \{e, e', ...\}$, associated with the vertex sets W and A, is composed of edges that are incident either with (1) two distinct vertices from A, or (2) a virtual vertex from W and a vertex from A. An edge incident with two distinct virtual vertices from W does not belong to the edge set E. Each virtual vertex is incident with, at least, one edge from E. A loop $l \in E$ is a special kind of edge incident only with one vertex, which must not be a virtual vertex. Loops can be used for description of free electron pairs. Diagrammatic representation of single components of the synthon is given in Fig. 1.

Definition 2.1 A *synthon* is an ordered triple

$$S(A) = (W, A, E) \tag{1}$$

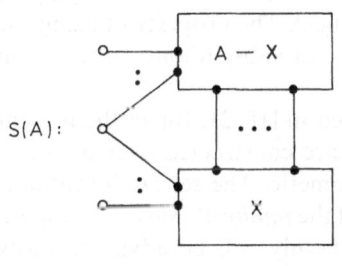

S(A):

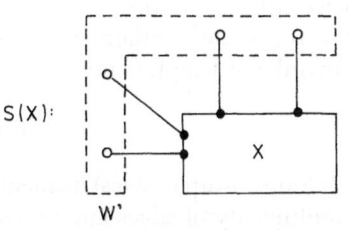

S(X):

Fig. 2. Convention for the construction of a subsynthon from a synthon. Let X be a proper subset of A, the synthon $S(A)$ is decomposed with respect to X into two parts. From the synthon $S(A)$ we construct a subsynthon $S(X) = (W', X, E')$ in such a way that its edge set is enlarged in a new edge set E' by those edges of $S(A)$ that are simultaneously incident with a vertex from X and either a virtual vertex or a vertex from $A \setminus X$. The used vertices from $A \setminus X$ are substituted by new virtual vertices, and the virtual vertices of both types (original and added) are assembled in the set W'. We say that the constructed subsynthon $S(X)$ is induced by the vertex subset X

127

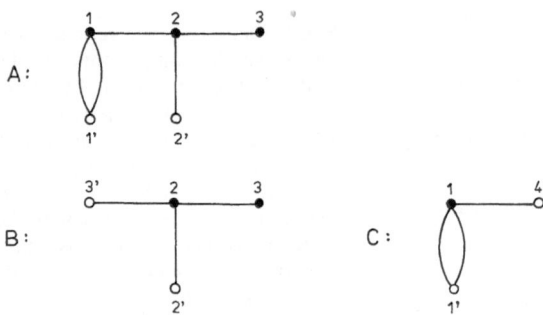

A:

B:

C:

Fig. 3. Examples of subsynthons (*B* and *C*) created from the synthon *A*. The subsynthon *B* is obtained from the synthon *A* specifying $X = \{2, 3\}$ whereas the subsynthon *C* for $X = \{1\}$

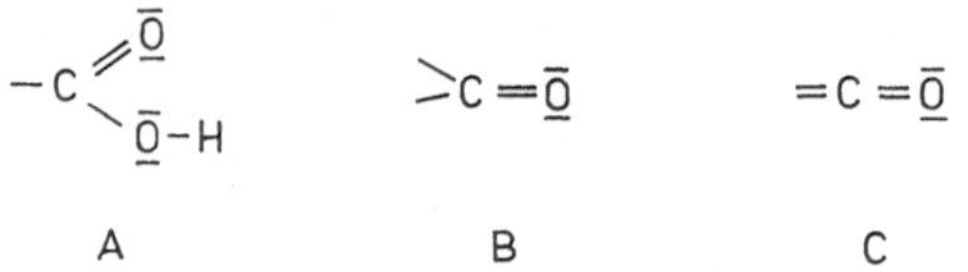

$$-C \overset{\overline{\overline{O}}}{\underset{\overline{\underline{O}}-H}{\diagdown}}$$

$$>C = \overline{\overline{\underline{O}}}$$

$$=C = \overline{\overline{\underline{O}}}$$

A   B   C

Fig. 4. Three structural fragments corresponding to the notion of synthon, the bond lines which stick out are conventionally terminated by "virtual" atoms (vertices). We see that the synthon *B* is a subsynthon of *A*, but that synthon *C* is not a subsynthon of *A*

where *W* is a vertex set composed of virtual vertices, *A* is a vertex set (non empty) composed of vertices (nonvirtual), and *E* is edge set associated with the vertex sets *W* and *A*.

The notion of a subsynthon of a synthon is defined as for subgraph of a graph [26]. A *subsynthon* is constructed from a given synthon $S(A) = (W, A, E)$ applying the rules of the following convention outlined in Fig. 2. The property of being the subsynthon is denoted by $S(X) \subseteq S(A)$. Examples of subsynthons are given in Figs. 3 and 4.

The notion of *isomerism* for synthons was defined in [18, 21, 16], in the present communication we shall study the synthons that are constructed over the same vertex set *A*. Therefore, they are automatically isomeric. The set of all synthons (nonisomorphic) constructed over the set *A* is called the *family of isomeric synthons*, and is denoted by $\mathscr{F}(A)$. The synthons from a family will be advantageously classified in our forthcoming considerations as *stable*, *unstable*, and *forbidden*. This will be done by making use of the concept of *valence states of vertices*.

In general, the valence state of a vertex $v \in A$ from the synthon $S(A)$ is determined after Ugi et al. [27] (cf. also Ref. [28]) as an ordered 4-tuple of integers

$$vs[v] = (n_0, n_1, n_2, n_3), \tag{2}$$

where n_0 (n_i, for $1 < i < 3$) is equal to the number of loops (*i*-tuple edges) incident with the vertex *v*. (We remember that the highest multiplicity of edges and loops is three.) The family $\mathscr{F}(A)$ may be divided into three disjoint subsets composed of the so-called *stable, unstable,* and *forbidden* valence states. The classification of a valence state under one of those three groups can be fitted to the demands of chemists.

Definition 2.2

1. The synthon $S(A)$ is *stable* if and only if the valence states of all its vertices are stable.
2. The synthon $S(A)$ is *unstable* if and only if its certain vertex is in an unstable valence state and contains no vertex in forbidden valence state.
3. The synthon $S(A)$ is *forbidden* if and only if at least one its vertex is in a forbidden valence state.

The above introduced classification scheme of synthons will be of great importance for the elaboration of effective heuristics to reduce the enormous number of synthons from the family $\mathscr{F}(A)$, where A is a given vertex set. Forbidden synthons are removed from the family $\mathscr{F}(A)$.

2.2 Chemical Transformation

Let us consider two distinct synthons $S_1(A) = (W_1, A, E_1)$ and $S_2(A) = (W_2, A, E_2)$ from the family $\mathscr{F}(A)$. These synthons may be formally related by a nonsymmetric relation called the *chemical transformation,*

$$S_1(A) \Rightarrow S_2(A), \tag{3}$$

where $S_1(A)$ $(S_2(A))$ is called the *educt (product)* synthon. Since this transformation "maps" an educt synthon onto a product synthon which is isomeric to the former one, the chemical transformation (3) may be formally understood as an *isomerization process.*

Each chemical transformation of two preselected synthons $S_1(A)$ and $S_2(A)$ will be represented by the so called *reaction graph* [16, 29] denoted by G_R.

The reaction graph is

$$G_R = (W_R, A_R, E_R, \psi, \{-1, +1\}), \tag{4}$$

where the edge and loop sets are equal to the disjunctive sum $(A + B = A \cup B \setminus A \cap B)$ [29] of the respective sets in $S_1(A)$ and $S_2(A)$,

$$E_R = E_1 + E_2, \tag{5}$$

The mapping $\psi : E_R \rightarrow \{-1, +1\}$ evaluates the edges and loops by $+1/-1$,

$$\psi(e) = \begin{cases} +1, & \text{for} \quad e \in E_2 \text{ (formed edge)}, \\ -1, & \text{for} \quad e \in E_1 \text{ (deleted edge)}, \end{cases} \tag{6}$$

The vertex set $A_R \subseteq A$ is composed of vertices incident with edges from E_R.

Unfortunately, the mapping of virtual vertices is not unambiguous due to the fact that the virtual vertex sets W_1 and W_2 of $S_1(A)$ and $S_2(A)$, respectively, are

Eva Hladka et al.

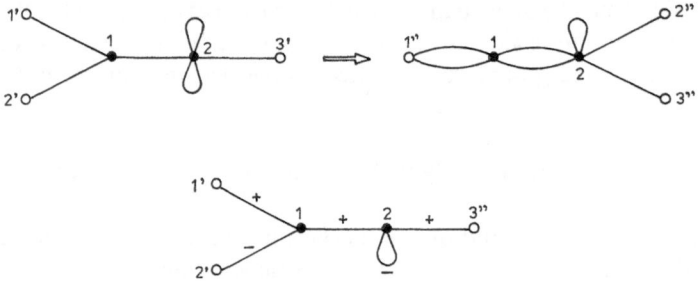

Fig. 5. An example of the construction of a reaction graph for a given chemical transformation displayed in the first row of figure. The resulting reaction graph G_R is given in the second row of the figure

in general different. Therefore, the edges of reaction graphs incident with one vertex and one virtual vertex are also not unambiguously determined. In order to overcome this formal difficulty the approach suggested by Koča [18, 21], based on the maximal common subgraph is applied. Therefore, there should be a 1–1 correspondence between virtual vertices of both synthons. An example of the construction of a reaction graph is given in Fig. 5.

2.3 Reaction Distance

The reaction distance [16, 18, 21, 25] between two isomeric synthons $S_1(A)$ and $S_2(A)$ will be used as a proper tool for the construction of reaction graphs [29]. The reaction graph obtained corresponds to the minimal number of the so-called *elementary chemical transformations*, the number of which determines the reaction distance between the synthons $S_1(A)$ and $S_2(A)$.

Let us consider a chemical transformation of the synthons $S_1(A)$ and $S_2(A)$,

$$S_1(A) \Rightarrow S_2(A), \tag{7}$$

this transformation is called the *elementary chemical transformation* for the following special cases:

1. The synthon $S_1(A)$ contains an edge $[v, v']$, the elementary transformation $\alpha^{vv'}$ "dissociates" the edge $[v, v']$ on a loop $[v', v']$, see Fig. 6. The resulting synthon $S_2(A)$ is determined by

$$\alpha^{vv'} : S_2(A) = (W_1, A, (E_1 \setminus \{[v, v']\}) \cup \{[v', v']\}). \tag{8}$$

2. The synthon $S_1(A)$ contains an edge $[v, w]$, the elementary transformations α^{vw} and α^{wv} are determined in Fig. 6. The resulting synthon $S_2(A)$ is determined by

$$\alpha^{vw} : S_2(A) = (W_1 \setminus \{w\}, A, E_1 \setminus \{[v, w]\}), \tag{9}$$

$$\alpha^{wv} : S_2(A) = (W_1 \setminus \{w\}, A, (E_1 \setminus \{[v, w]\}) \cup \{[v, v]\}). \tag{10}$$

130

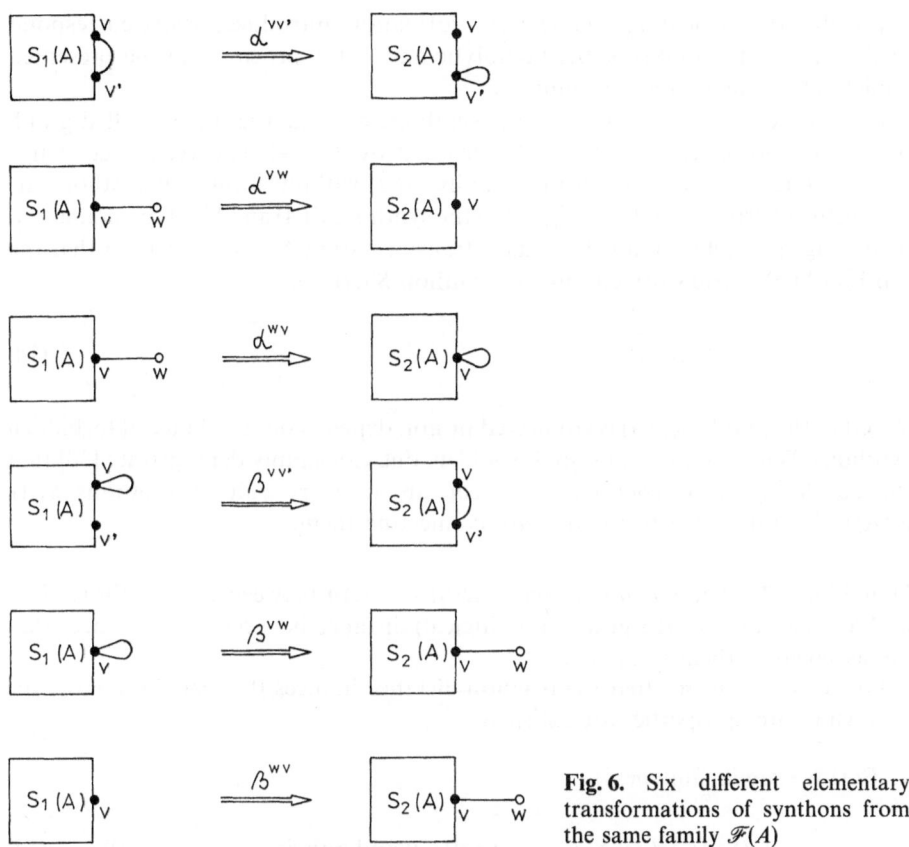

Fig. 6. Six different elementary transformations of synthons from the same family $\mathscr{F}(A)$

3. The synthon $S_1(A)$ contains a loop $[v, v]$, the elementary transformations $\beta^{vv'}$ and β^{vw} (where $v' \in A$ and $w \notin W_1$) "associate" the loop on an edge $[v, v']$ and $[v, w]$, respectively, see Fig. 6. The resulting synthons are

$$\beta^{vv'} : S_2(A) = (W_1, A, (E_1 \cup \{[v, v']\}) \setminus \{[v, v]\})\,, \tag{11}$$

$$\beta^{vw} : S_2(A) = (W_1 \cup \{w\}, A, (E_1 \cup \{[v, w]\}) \setminus \{[v, v]\})\,. \tag{12}$$

4. The synthon $S_1(A)$ contains a vertex $v \in A$, the elementary transformation β^{wv} is schematically determined in Fig. 6. The resulting synthon $S_2(A)$ is

$$\beta^{wv} : S_2(A) = (W_1 \cup \{w\}, A, E_1 \cup \{[v, w]\})\,. \tag{13}$$

We see that $\beta^{v'v}$ (β^{vw} and β^{wv}) is a retrotransformation with respect to $\alpha^{vv'}$ (α^{wv} and α^{vw}). They represent the most elementary processes involved in mechanistic considerations of chemical transformations of a synthon into another one. We have to emphasize that, in order to keep our forthcoming considerations as simple as possible, the elementary chemical transformations introduced above correspond to the heterolytic processes, the homolytic and redox processes in the presented model are not taken into account.

For a fixed family $\mathscr{F}(A)$ of isomeric synthons we construct the so-called graph of reaction distances [18, 21, 16, 25] denoted by $\mathscr{G}_{RD}(A)$. The vertex set of this graph is formally identical with the family $\mathscr{F}(A)$ without forbidden synthons, its two distinct vertices v and v', assigned to the synthons $S(A)$ and $S'(A)$, are connected by an edge $[v, v']$ if such an elementary transformation $\xi = \alpha, \beta$ exists so that the synthon $S(A)$ is transformed into the synthon $S'(A)$, i.e.

$$S(A) \overset{\xi}{\Rightarrow} S'(A) \tag{14}$$

Whether the graph $\mathscr{G}_{RD}(A)$ is connected or not, depends on the choice of forbidden synthons. When no synthons are forbidden, one can simply demonstrate [30] that the graph $\mathscr{G}_{RD}(A)$ is *connected*, between an arbitrary pair of synthons $S(A)$, $S'(A) \in \mathscr{F}(A)$ there exists a finite path connecting them.

Definition 2.3 The *reaction distance* $D(S(A), S'(A))$ between two synthons $S(A)$ and $S'(A)$ is equal to the graph (topological) distance between those vertices that are assigned to them in $\mathscr{G}_{RD}(A)$.

One can easy to see that the reaction distance induces the reaction metric, the following three properties are satisfied:

1. Positive semidefiniteness

$$D(S(A), S'(A)) > 0 . \qquad (= 0 \quad \text{if and only if} \quad S(A) = S'(A)) . \tag{15}$$

2. Symmetry

$$D(S(A), S'(A)) = D(S'(A), S(A)) . \tag{16}$$

3. Triangle inequality

$$D(S(A), S'(A)) + D(S'(A), S''(A)) \geq D(S(A), S''(A)) . \tag{17}$$

where the equality is satisfied if and only if the synthon $S'(A)$ lies on a shortest path connecting the synthons $S(A)$ and $S''(A)$.

The proof of these properties is obvious, it immediately follows from the properties of the graph distance.

In order to illustrate the above considerations, let us study a subfamily of $\mathscr{F}(A)$, where $|A| = 1$ and $0 \leq |E| \leq 3$, it contains the 14 synthons displayed in Fig. 7. For instance, the synthon $S_5(A)$ may be transformed by four admissible elementary transformations,

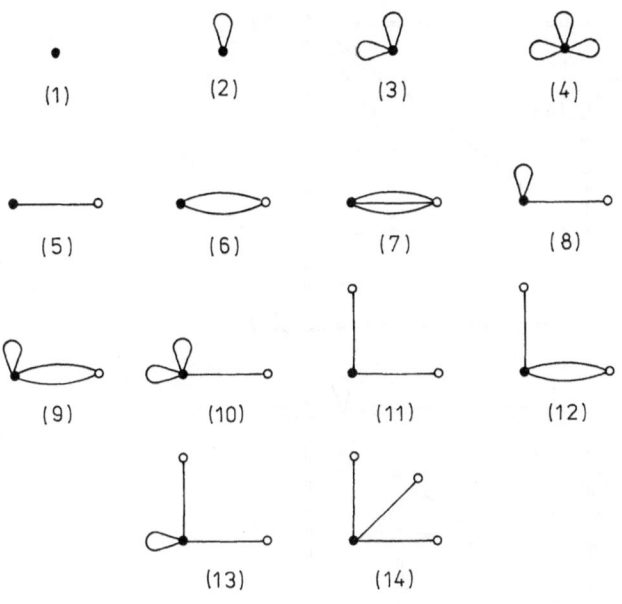

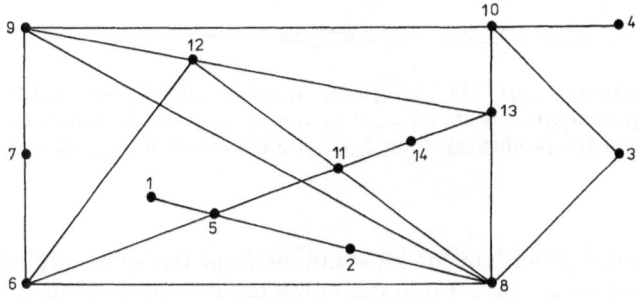

Fig. 7. An example of subfamily $\mathscr{F}(A)$, where $|A| = 1$ and $0 \leq |E| \leq 3$, it contains 14 synthons indexed by 1 to 14. The corresponding graph of reaction distances $\mathscr{G}_{RD}(A)$ is composed of 14 vertices (synthons)

$$S_5(A) \overset{\alpha^{11'}}{\Rightarrow} S_1(A), \tag{18}$$

$$S_5(A) \overset{\alpha^{1'1}}{\Rightarrow} S_2(A), \tag{19}$$

$$S_5(A) \overset{\beta^{2'1}}{\Rightarrow} S_{11}(A), \tag{20}$$

$$S_5(A) \overset{\beta^{1'1}}{\Rightarrow} S_6(A). \tag{21}$$

Hence, in the graph $\mathscr{G}_{RD}(A)$ the vertex assigned to the synthon $S_5(A)$ is connected by edges with the vertices corresponding to the synthons $S_1(A)$, $S_2(A)$, $S_6(A)$, and

133

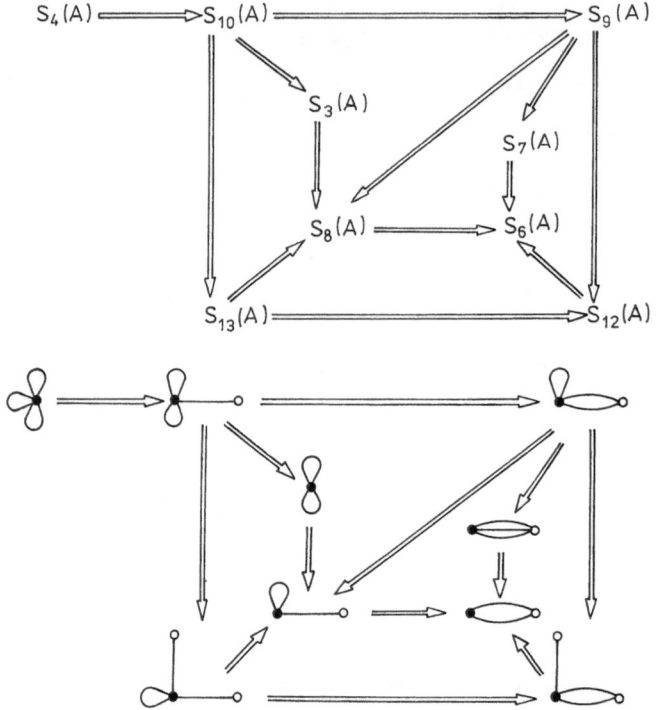

Fig. 8. Two versions of a "reaction network" [24, 71, 72] for the transformation $S_4(A) \Rightarrow S_6(A)$, in the second version the synthon symbols are replaced by their graphical representations. The educt synthon $S_4(A)$ and the product synthon $S_6(A)$ are connected by six different oriented paths

$S_{11}(A)$. Applying the similar procedure for all synthons from the subfamily of $\mathscr{F}(A)$ we arrive at $\mathscr{G}_{RD}(A)$ given in Fig. 7. From this graph one can simply evaluate the reaction distance for an arbitrary pair of synthons, e.g. $D(S_1(A), S_4(A)) = 6$ and $D(S_4(A), S_6(A)) = 4$. Let us study the transformation $S_4(A) \Rightarrow S_6(A)$, it can be decomposed in the six sequences of elementary transformations displayed in Fig. 8. Applying the chemical terminology, the transformation $S_4(A) \Rightarrow S_6(A)$ has six different mechanisms expressed by the following sequences of four elementary transformations

$$S_4(A) \Rightarrow S_{10}(A) \Rightarrow S_9(A) \Rightarrow S_7(A) \Rightarrow S_6(A),$$

$$S_4(A) \Rightarrow S_{10}(A) \Rightarrow S_9(A) \Rightarrow S_{12}(A) \Rightarrow S_6(A),$$

$$S_4(A) \Rightarrow S_{10}(A) \Rightarrow S_{13}(A) \Rightarrow S_{12}(A) \Rightarrow S_6(A),$$

$$S_4(A) \Rightarrow S_{10}(A) \Rightarrow S_{13}(A) \Rightarrow S_8(A) \Rightarrow S_6(A),$$

$$S_4(A) \Rightarrow S_{10}(A) \Rightarrow S_3(A) \Rightarrow S_8(A) \Rightarrow S_6(A),$$

$$S_4(A) \Rightarrow S_{10}(A) \Rightarrow S_9(A) \Rightarrow S_8(A) \Rightarrow S_6(A).$$

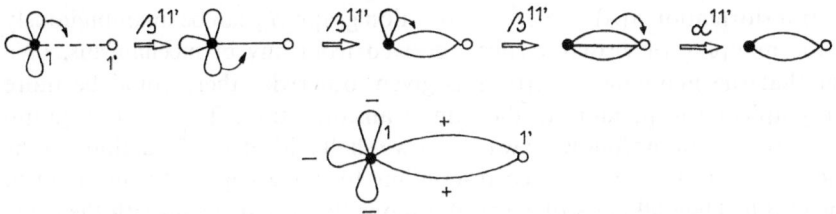

Fig. 9. A diagrammatic interpretation of the first mechanism together with the produced reaction graph corresponding to the chemical transformation $S_4(A) \Rightarrow S_6(A)$ (see Fig. 8)

For instance, a diagrammatic interpretation of the first mechanism together with the reaction graph produced is given in Fig. 9.

The shortest paths are unambiguously represented by sequences of k elementary transformations $\xi_1, \xi_2, \ldots, \xi_k$, the corresponding mechanisms of $S(A) \Rightarrow S'(A)$ are then determined by

$$S(A) \overset{\xi_1}{\Rightarrow} S_1(A) \overset{\xi_2}{\Rightarrow} \ldots \overset{\xi_{k-1}}{\Rightarrow} S_{k-1}(A) \overset{\xi_k}{\Rightarrow} S'(A) \tag{22}$$

For instance, the sequences of elementary transformations of the mechanisms listed in the above example are

1st mechanism:	$\beta^{11'}$	$\beta^{11'}$	$\beta^{11'}$	$\alpha^{11'}$,
2nd mechanism:	$\beta^{11'}$	$\beta^{11'}$	$\beta^{12'}$	$\alpha^{12'}$,
3rd mechanism:	$\beta^{11'}$	$\beta^{12'}$	$\beta^{11'}$	$\alpha^{12'}$,
4th mechanism:	$\beta^{11'}$	$\beta^{12'}$	$\alpha^{12'}$	$\beta^{11'}$,
5th mechanism:	$\beta^{11'}$	$\alpha^{11'}$	$\beta^{11'}$	$\beta^{11'}$,
6th mechanism:	$\beta^{11'}$	$\beta^{11'}$	$\alpha^{11'}$	$\beta^{11'}$.

The reaction graphs corresponding to these mechanisms are displayed in Fig. 10.

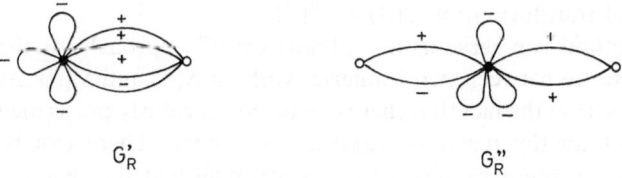

$$G'_R \qquad\qquad\qquad G''_R$$

Fig. 10. The first (second) graph corresponds to the 1st, 5th, and 6th (2nd to 4th) mechanisms. These graphs were constructed by adding edges and loops of reaction graphs of all elementary transformations of a mechanism into one graph. The corresponding (reduced) reaction graph is given in Fig. 9

For a transformation $S(A) \Rightarrow S'(A)$ the reaction graph G_R can be unambiguously (up to the mapping of virtual vertices) derived from any of mechanisms. (We consider that the mapping of vertices is given, otherwise there could be more reaction graphs corresponding to the same transformation). The reaction graph can be constructed in the following way: first we shall add edges of reaction graphs of all elementary transformations corresponding to single steps of the mechanism into one graph. Then all pairs of edges and loops that are incident with the same vertices but of opposite evaluation ± 1 are deleted from the graph. Finally, the occurred isolated vertices are also deleted. The graph obtained is identical with the reaction graph G_R of the transformation $S(A) \Rightarrow S'(A)$. These considerations are summarized in the form of the following property: Each reaction graph $G_R = (W_R, A_R, E_R, \psi, \{+1, -1\})$ satisfies the following inequality:

$$|E_R| + |L_R| \leq 2D(S(A), S'(A)). \tag{23}$$

An inverse problem to the one above consists of a construction of a sequence of elementary transformations from the given reaction graph. Let us consider a transformation $S(A) \Rightarrow S'(A)$, where $D(S(A), S'(A)) = k > 0$. The transformation may be "algebraized" [28] by making use of the concept of reaction graph G_R,

$$S(A) + G_R = S'(A), \tag{24}$$

where the binary operation '+' corresponds to the disjunctive sum of sets. We try to decompose the reaction graph G_R into a sum of elementary reaction graphs corresponding to elementary transformations composed of a minimal number of terms

$$G_R = \sum_{i=1}^{l} G_R^{(i)}. \tag{25}$$

The minimal value of l corresponds to an upper bound of the reaction distance,

$$k = D(S(A), S'(A)) \leq \min l. \tag{26}$$

All admissible ways of decomposition of G_R into k elementary reaction graphs express the mechanisms of transformation $S(A) \Rightarrow S'(A)$.

The above outlined method can serve as an "almost exact" approach for the evaluation of reaction distance between two isomeric synthons $S(A)$ and $S'(A)$. Its "almost exactness" follows from the fact that there can be no previously prescribed mapping of vertices and than the reaction graph is not unique. There can be constructed corresponding reaction graphs for every mapping and the reaction distance should be obtained as a minimum of minimal coverings of those reaction graphs.

In our communication [25] we suggested the so-called *bilateral approach*, it provides a general method for the evaluation of reaction distance. Unfortunately, its numerical efficiency fast decreases (a combinatorial explosion) for pairs of synthons with greater reaction distance. Therefore, for actual applications of the present theory the difficulties with an efficient evaluation of the reaction distances are simply overcame by the approach of minimal covering of the reaction graph. The reaction graph is constructed on the base of given mapping of vertices and/or on the base of the maximal common subgraph. The resulting reaction distance may be then in some special cases greater than its exact value. The inequalities (23) and (26) provide the lower and upper bounds for the reaction distance,

$$\tfrac{1}{2}\left(|E_R| + |L_R|\right) \le D(S(A), S'(A)) \le \min l. \tag{27}$$

2.4 Synthetic Precursors/Successors of a Synthon

The purpose of this subsection is to define for a given synthon $S(A)$ its synthetic precursor and/or successor (SPS) in such a way that this term will be closely related to the reasoning of organic chemists. An organic synthesis of a product compound M' from an educt compound M is usually realized not in one step but it is carried out through a multistage synthetic procedure, schematically

$$M^{(0)} = M \Rightarrow M^{(1)} \Rightarrow M^{(2)} \Rightarrow \ldots \Rightarrow M^{(n)} = M'. \tag{28}$$

An intermediate compound $M^{(i)}$ (for $1 \le i \le n - 1$) corresponds to a stable (or relatively stable) compound that is closely related to its precursor $M^{(i-1)}$. Moreover, the chemical transformation $M^{(i-1)} \Rightarrow M^{(i)}$ is usually realized [31] via a mechanism composed of a sequence of non-branching simple dissociations and associations of bonds and lone-electron pairs, respectively. Loosely speaking, such a mechanism corresponds to a "push-pull" [31] non-branched flow of valence electrons through a skeleton of $M^{(i-1)}$.

Let us consider a pair of distinct synthons $S(A)$ and $S'(A)$ taken from the family $\mathscr{F}(A)$ of isomeric synthons, their reaction distance is equal to $k > 1$, i.e. $D(S(A), S'(A)) = k$. The shortest paths in $\mathscr{G}_{RD}(A)$, connecting the vertices $S(A)$ and $S'(A)$ are denoted by $P_1, P_2, \ldots, P_r$; they are assembled at a set $\mathscr{P}(S(A), S'(A))$. Each path of this set is composed of k edges and $k + 1$ vertices, where the initial vertex and the terminal vertex correspond to the synthons $S(A)$ and $S'(A)$, respectively.

Definition 2.4 A stable neighborhood of the synthon $S(A)$, denoted by $\hat{\mathfrak{S}}(S(A))$, is a set composed of stable synthons $S'(A)$ such that the internal vertices from each path of $\mathscr{P}(S(A), S'(A))$ correspond merely to unstable synthons.

We shall assume a priori that the family $\mathscr{F}(A)$ has at least two stable synthons and the graph $\mathscr{G}_{RD}(A)$ is connected. As a consequence each synthon $S(A)$ has a nonempty stable neighborhood, i.e. $\hat{\mathfrak{S}}(S(A)) \neq \emptyset$.

The notion of a stable neighborhood of a synthon will be illustrated by using the subfamily $\mathscr{F}(A)$ displayed in Fig. 7, the corresponding graph of reaction distances is given in Fig. 8. Now we will assume that the synthons $S_3(A)$, $S_9(A)$, and $S_{13}(A)$ are the only stable synthons (they are denoted by encircled vertices), see

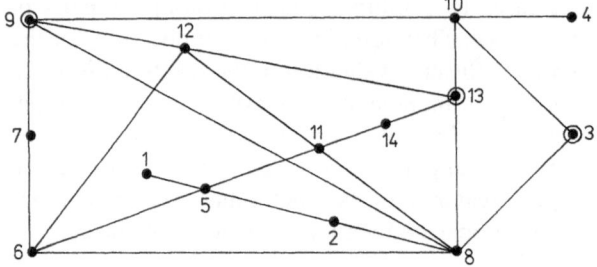

Fig. 11. Modified graph of reaction distances $\mathscr{G}_{RD}(A)$ from Fig. 7, now the synthons indexed by 3, 9, and 13 are declared as the only stable ones *(encircled vertices)*

Fig. 11. For instance, let us consider the following pair of synthons: $S(A) = S_{12}(A)$ and $S'(A) = S_3(A)$, where $S_3(A)$ is a stable synthon. The reaction distance between these synthons is $D(S(A), S'(A)) = 3$. There exist six shortest paths from $S_{12}(A)$ to $S_3(A)$,

$$
\begin{aligned}
P_1: &\quad 12\text{–}13\text{–}10\text{–}3\,, \\
P_2: &\quad 12\text{–}13\text{–}8\text{–}3\,, \\
P_3: &\quad 12\text{–}9\text{–}10\text{–}3\,, \\
P_4: &\quad 12\text{–}6\text{–}8\text{–}3\,, \\
P_5: &\quad 12\text{–}9\text{–}8\text{–}3\,, \\
P_6: &\quad 12\text{–}11\text{–}8\text{–}3\,.
\end{aligned}
$$

Since the paths P_1 and P_2 contain as internal vertex the stable synthon $S_{13}(A)$, the stable synthon $S_3(A)$ is not an element of the stable neighborhood of the synthon $S_{12}(A)$. The stable neighborhood of $S_{12}(A)$ is $\hat{\mathfrak{S}}(S_{12}(A)) = \{S_9(A), S_{13}(A)\}$. In other words, a stable neighbourhood of a synthon $S(A)$ is composed only of those synthons $S'(A)$, for which there exists no shortest path between $S(A)$ and $S'(A)$ in graph $\mathscr{G}_{RD}(A)$ containing as an intermediate some stable synthon.

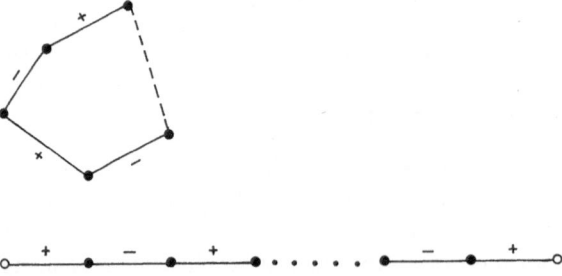

Fig. 12. Two different "structures" of reaction graphs used in definition 2.5. The first one (cyclic) does not contain any virtual vertices while the second one (linear) contains two virtual vertices

Definition 2.5 A stable synthon $S'(A)$ is called *synthon precursor/successor* (SPS) of the stable synthon $S(A)$ iff the reaction graph G_R of the transformation $S(A) \Rightarrow S'(A)$ is of a cyclic or linear form, see Fig. 12. Both forms must have alternating evaluation of edges by $+/-$ and the linear graph must have terminal virtual vertices. The set of all possible SPS of the synthon $S(A)$ is denoted by $\mathfrak{S}(S(A))$.

Supposition 2.1 Each synthon $S(A)$ has a nonempty set of SPS i.e. $\mathfrak{S}(S(A)) \neq \emptyset$.

In order to keep this very important supposition in its correctness we have to postulate that the family $\mathscr{F}(A)$ and the graph $\mathscr{G}_{GD}(A)$ are of sufficiently structurally diverse nature.

Supposition 2.2 For a given stable synthon $S(A)$ we have

$$\mathfrak{S}(S(A)) \subseteq \hat{\mathfrak{S}}(S(A)), \tag{29}$$

that is each SPS of $S(A)$ belongs to the stable neighborhood of $S(A)$.

Let us consider a stable synthon $S'(A) \in \mathfrak{S}(S(A))$, the transformation $S(A) \Rightarrow S'(A)$ is represented by a reaction graph either of the linear or cyclic form. In order to verify that the synthon $S'(A)$ belongs to the stable neighborhood of the stable synthon $S(A)$ it is sufficient to verify that each connected subgraph of G_R produces the unstable synthon. As this subgraph would be of linear form, at least one nonvirtual atom would gain or loose one valence electron and therefore its valence state would be unstable and the synthon $S'(A)$ produced should not be stable.

For a moment, let us come back to the organic chemistry. An organic reaction, formally treated as the transformation $S(A) \Rightarrow S'(A)$, where $S(A)$ and $S'(A)$ are stable synthons, is carried out as a sequence of transformations (not necessary elementary transformations),

$$S^{(0)}(A) = S(A) \Rightarrow S^{(1)}(A) \Rightarrow \ldots \Rightarrow S^{(n)}(A) = S'(A), \tag{30}$$

where the i-th intermediate stable synthon $S^{(i)}(A)$ simultaneously belongs to the sets $\mathfrak{S}(S^{(i-1)}(A))$ and $\mathfrak{S}(S^{(i+1)}(A))$. From the point of view of $S^{(i-1)}(A)$ the synthon $S^{(i)}(A)$ is a successor whereas from the point of view of $S^{(i+1)}(A)$ this synthon is a precursor. We see that the concept of SPS is closely related to reasoning of organic chemists.

Since the number of synthons in $\mathfrak{S}(S(A))$ is usually enormous, we turn our attention to its reduction to a substantially smaller subset still containing all synthetically important SPS of the synthon $S(A)$. Let X be a subset of the vertex set A, it will be called the *reaction center*. A subset of $\mathfrak{S}(S(A))$ with respect to the reaction center is determined as follows.

Definition 2.6 For a given synthon $S(A) = (W, A, E)$ a subset $\mathfrak{S}(S(A/X)) \subset \mathfrak{S}(S(A))$, introduced with respect to the reaction center X, is composed of those synthons $S'(A) = (W, A, E') \in \mathfrak{S}(S(A))$ that satisfy

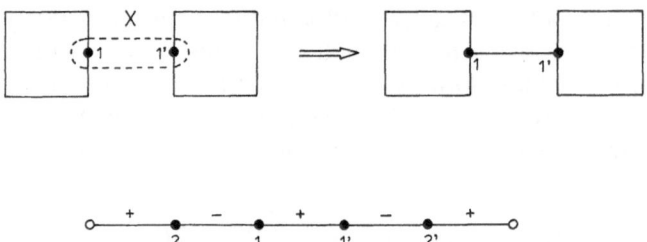

Fig. 13. Schematic illustration of definition 2.6. In the first row a chemical transformation is displayed in which a synthon with two nonincident vertices from the reaction set X is transformed into another synthon with already incident vertices which were initially nonincident. In the second row a possible reaction graph is displayed

1. for each pair of distinct vertices $v \in A$ and $v' \in A \setminus X$ we have

$$[v, v'] \notin E \Rightarrow [v, v'] \notin E' ,$$

2. the vertex set of G_R assigned to $S(A) \Rightarrow S'(A)$ contains, at least, one vertex from the reaction center X, i.e.

$$\exists v \in A_R : v \in X .$$

It means that a stable synthon $S'(A) \in \mathfrak{S}(S(A/X))$ is restricted such that if a vertex $v \in X$ (or $v \in A \setminus X$) and another one $v' \in A \setminus X$, where $v \neq v'$, are not connected by an edge in $S(A)$, then these vertices are not also connected by an edge in $S'(A)$. In other words, going from the stable synthon $S(A)$ to another stable synthon $S'(A) \in \mathfrak{S}(S(A/X))$ in the graph $\mathscr{G}_{RD}(A)$, new edges (with unit multiplicity) may be formed only inside the reaction center, outside the reaction center the process of formation of new edges (i.e. bonds) is forbidden. This implies that if the synthon $S(A)$ is disconnected (i.e. it contains at least two components) and the resulting synthon $S'(A)$ contains smaller number of the components than $S(A)$ (it may be, in extreme cases, connected), the reaction center X should be composed of vertices that belong to different components of the synthon $S(A)$, see Fig. 13. The possible corresponding graph G_R is also displayed in Fig. 13. It has one edge evaluated by the symbol '+' between vertices from the set X, this edge is incident with vertices that belong to different components of the synthon $S(A)$.

For an actual application of the present theory, it is often very important to keep some fragments of the synthon $S(A)$ outside the reaction center X intact during the applied chemical transformation. For instance, for synthesizing a biologically active compound, the intact skeleton would be responsible for its target activity. Such a requirement can be implemented in our theory by making use of the so-called *intact set* $\bar{X}$ of vertices, where $\bar{X} \subset A$ and $X \cap \bar{X} = \emptyset$.

Definition 2.7 A subset $\mathfrak{S}(S(A/X/\bar{X})) \subset \mathfrak{S}(S(A/X))$ for a given reaction center X and an intact set $\bar{X}$ is composed of those stable synthons $S'(A) \in \mathfrak{S}(S(A/X))$

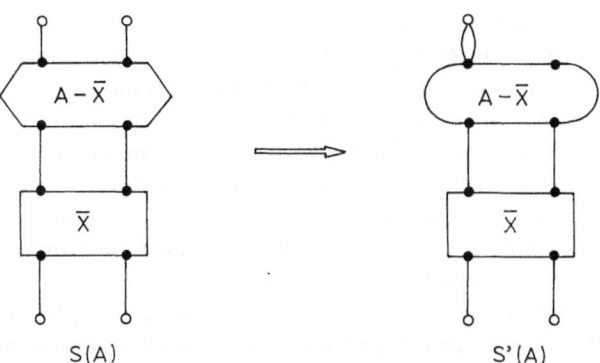

Fig. 14. Schematic representation of single terms used in definition 2.7. Synthon $S'(A) \in \mathfrak{S}(S(A/X/\bar{X}))$ is different from the synthon $S(A)$ only in the upper block containing the vertices from $A \setminus \bar{X}$. The bottom blocks represent the subsynthon $S(\bar{X})$ and $S'(\bar{X})$, respectively, which are mutually isomorphic. In other words, the subsynthons $S(\bar{X})$ and $S'(\bar{X})$ remain intact during the transformation $S(A) \Rightarrow S'(A)$

such that the graph G_R assigned to $S(A) \Rightarrow S'(A)$ does not affect the edges that are incident simultaneously with two distinct vertices of the intact set $\bar{X}$.

In order to illustrate the above definition, let us consider a synthon $S'(A)$ such that $S'(A) \subset \mathfrak{S}(S(A/X/\bar{X}))$, the transformation $S(A) \Rightarrow S'(A)$ can be schematically represented in Fig. 14.

These synthons are different only in the upper blocks containing the vertices from $A \setminus \bar{X}$, the bottom blocks induce the subsynthons $S(\bar{X})$ and $S'(\bar{X})$, respectively, that are mutually isomorphic, $S(\bar{X}) \approx S'(\bar{X})$. In other words, the subsynthons $S(\bar{X})$ and $S'(\bar{X})$ remain intact during the transformation $S(A) \Rightarrow S'(A)$.

Following the above definitions we have

$$\mathfrak{S}(S(A/X/\bar{X})) \subseteq \mathfrak{S}(S(A/X)) \subseteq \mathfrak{S}(S(A)) \subseteq \hat{\mathfrak{S}}(S(A)), \tag{31}$$

$$\mathfrak{S}(S(A/X/\bar{X} = \emptyset)) = \mathfrak{S}(S(A/X)), \tag{32}$$

$$\mathfrak{S}(S(A/X = A)) = \mathfrak{S}(S(A)). \tag{33}$$

Equation (31) is obvious, it immediately follows from the definitions of the sets $\mathfrak{S}(S(A/X/\bar{X}))$, $\mathfrak{S}(S(A/X))$, and $\mathfrak{S}(S(A))$. Equation (32) means that if an intact set X is empty, then the subset subset $\mathfrak{S}(S(A/X/\bar{X}))$ is identical with the subset $\mathfrak{S}(S(A/X))$. Finally, Eq. (33) means that if a reaction center X is extended over whole set A, then the subset $\mathfrak{S}(S(A/X))$ is equal to the set $\mathfrak{S}(S(A))$.

3 The Mathematical Model as the Formal Basis of the Program PEGAS

Chemical structural formulae, which are the basis of the chemical pictorial language, are also the universal tool for a description of topological information about the molecule. These "pictures" explicitly contain qualitative information about atoms in the molecule as well as facts about chemical bonds, i.e. they inform which couples of atoms are so close to each other that it can be considered that there is a chemical bond (covalent or ionic). The structural formula also has implicit information value which can be surprisingly high. It can be shown, for example, by application of various topological indices for the prediction of physico-chemical or biological properties [32–34]. At the same time, the structural formula is the most often used tool for abstract manipulation with the molecule. The structural formula, i.e. a topological expression of the molecule, is also a basis of our model. The model has been formulated in two, in principle very similar, mathematical forms. The former, described in the previous section, is expressed by graph-theoretical language. The latter is based on a linear algebra approach, and it is here discussed in the fashion which is implemented in the program PEGAS.

The individual atom and its electronic configuration, the so-called valence state of atom, is the elementary entity of the model philosophy. The modeling of valence states of atoms and their interconversions is then extended to larger systems, the so-called synthons (see Sect. 3.5).

3.1 Valence States of Atoms

The notion of the Valence State of an atom is based on that of Pauling [35] and Van Vleck [36] called "atom in molecule". When considering only "integral chemistry", where the multiplicity of bonds is represented by positive integers, and moreover, postulating that the highest multiplicity of the chemical bond considered is three[1], the valence state of the atom can be described as a four-dimensional vector VVS [27, 37, 28, 38]:

$$VVS = (v_1, v_2, v_3, v_4) \qquad (34)$$

where v_1 is number of free electrons, v_2, v_3, and v_4 is number of single, double and triple bonds, respectively.

Example 3.1 For the most frequent valence states of carbon and oxygen atom we have:

$$-\overset{|}{\underset{|}{C}}- : (0, 4, 0, 0); \quad =\overset{|}{C}- : (0, 2, 1, 0); \quad \equiv C- : (0, 1, 0, 1); \quad -\overline{O}- : (4, 2, 0, 0);$$

$$=\overline{O}: (4, 0, 1, 0).$$

Within the frame of octet-chemistry, Eq. (34) implies:

$$v_1 + v_2 + 2v_3 + 3v_4 \leq 8 \qquad (35)$$

Since all $v_1, \ldots, v_4$ are non-negative integers, Eq. (35) results in 136 possible valence states of atoms within the frame of octet chemistry. All of them are, for a general atom X, shown in Table 1.

---

[1] In principle, the model is open to extension for considering bonds of multiplicity four and higher.

Table 1. Valence states of a general atom X and their VVS. All the valence states are considered in the frame of octet-chemistry. NSDT is number of n-electrons, single, double, and triple bonds

No.		NSDT	No.		NSDT	No.		NSDT	No.		NSDT
0	X	0 0 0 0	34	>X̲<	4 4 0 0	68	>Ẋ=	1 2 1 0	102	Ẋ≡	1 0 0 1
1	X–	0 1 0 0	35	Ẋ	5 0 0 0	69	≥Ẋ=	1 3 1 0	103	–Ẋ≡	1 1 0 1
2	–X–	0 2 0 0	36	Ẋ–	5 1 0 0	70	≥Ẋ<	1 4 1 0	104	>Ẋ≡	1 2 0 1
3	–X<	0 3 0 0	37	Ẋ<	5 2 0 0	71	≥Ẋ<	1 5 1 0	105	≥Ẋ≡	1 3 0 1
4	>X<	0 4 0 0	38	Ẋ<	5 3 0 0	72	–X̄=	2 1 1 0	106	≥Ẋ≡	1 4 0 1
5	>X≤	0 5 0 0	39	X̲	6 0 0 0	73	>X̄=	2 2 1 0	107	X̄≡	2 0 0 1
6	≥X≤	0 6 0 0	40	X̲–	6 1 0 0	74	≥X̄=	2 3 1 0	108	–X̄≡	2 1 0 1
7	≥X̣≤	0 7 0 0	41	X̲<	6 2 0 0	75	≥X̄<	2 4 1 0	109	>X̄≡	2 2 0 1
8	≥Ẍ≤	0 8 0 0	42	X̲·	7 0 0 0	76	–Ẋ=	3 1 1 0	110	≥X̄≡	2 3 0 1
9	Ẋ	1 0 0 0	43	X̲·	7 1 0 0	77	>Ẋ=	3 2 1 0	111	Ẋ≡	3 0 0 1
10	Ẋ–	1 1 0 0	44	X̲∣	8 0 0 0	78	≥Ẋ=	3 3 1 0	112	–Ẋ≡	3 1 0 1
11	–Ẋ–	1 2 0 0	45	X=	0 0 1 0	79	–X̄=	4 1 1 0	113	>Ẋ≡	3 2 0 1
12	–Ẋ<	1 3 0 0	46	=X=	0 0 2 0	80	>X̄=	4 2 1 0	114	X̄≡	4 0 0 1
13	>Ẋ<	1 4 0 0	47	=X≤	0 0 3 0	81	Ẋ<	5 1 1 0	115	–X̄≡	4 1 0 1
14	>Ẋ≤	1 5 0 0	48	≥X<	0 0 4 0	82	=X=	0 1 2 0	116	Ẋ≡	5 0 0 1
15	≥Ẋ≤	1 6 0 0	49	Ẋ=	1 0 1 0	83	=X=	0 2 2 0	117	=X≡	0 0 1 1
16	≥Ẋ≤	1 7 0 0	50	=Ẋ=	1 0 2 0	84	=X=	0 3 2 0	118	=X≡	0 1 1 1
17	X̄	2 0 0 0	51	=Ẋ≤	1 0 3 0	85	=X=	0 4 2 0	119	=X≡	0 2 1 1
18	X̄–	2 1 0 0	52	X̄=	2 0 1 0	86	=Ẋ=	1 1 2 0	120	=X≡	0 3 1 1
19	–X̄–	2 2 0 0	53	=X̄=	2 0 2 0	87	=Ẋ=	1 2 2 0	121	=X≡	1 0 1 1
20	–X̄<	2 3 0 0	54	=X̄≤	2 0 3 0	88	=Ẋ=	1 3 2 0	122	=X≡	1 1 1 1
21	>X̄<	2 4 0 0	55	Ẋ=	3 0 1 0	89	=X̄=	2 1 2 0	123	=Ẋ≡	1 2 1 1
22	>X̄≤	2 5 0 0	56	=Ẋ=	3 0 2 0	90	=X̄=	2 2 2 0	124	=X̄≡	2 0 1 1
23	≥X̄≤	2 6 0 0	57	X̄=	4 0 1 0	91	≥Ẋ=	3 1 2 0	125	=X̄≡	2 1 1 1
24	Ẋ	3 0 0 0	58	=X̄=	4 0 2 0	92	≥X=	0 1 3 0	126	=Ẋ≡	3 0 1 1
25	Ẋ–	3 1 0 0	59	·X̄=	5 0 1 0	93	≥X=	1 1 3 0	127	≥X≡	0 0 2 1
26	–Ẋ–	3 2 0 0	60	X̄=	6 0 1 0	94	≥X=	0 2 3 0	128	≥X≡	1 0 2 1
27	–Ẋ<	3 3 0 0	61	–X=	0 1 1 0	95	X≡	0 0 0 1	129	≥X≡	0 1 2 1
28	>Ẋ<	3 4 0 0	62	>X=	0 2 1 0	96	≡X≡	0 0 0 2	130	=X≡	0 1 0 2
29	>Ẋ≤	3 5 0 0	63	≥X=	0 3 1 0	97	–X≡	0 1 0 1	131	≡X≡	0 2 0 2
30	X̄	4 0 0 0	64	>X≤	0 4 1 0	98	>X≡	0 2 0 1	132	≡X≡	1 0 0 2
31	X̄–	4 1 0 0	65	≥X≤	0 5 1 0	99	≥X≡	0 3 0 1	133	≡X̄≡	2 0 0 2
32	–X̄–	4 2 0 0	66	≥X≤	0 6 1 0	100	≥X≡	0 4 0 1	134	≡Ẋ≡	1 1 0 2
33	–X̄<	4 3 0 0	67	–Ẋ=	1 1 1 0	101	>Ẋ≡	0 5 0 1	135	≡X≡	0 0 1 2

Valence states of individual atoms can be classified from various points of view. We will categorize them as un-charged, anions, cations, and radicals, or as stable and unstable. For some atoms, the valence states have been extracted from the literature, and they are shown in Table 2. Note that Table 2 is open to further extension. The notion of a stable valence state is used for atoms without a formal

Table 2. Valence states of selected chemical elements (part of the table introduced in Kratochvíl M, Koča J (1985) Chemistry in Graphs. Lachema, Brno (in Czech))

Symbols: ⊕ cation, ⊖ anion, ⊙ radical, ○ un-charged, ⊗ un-specified

Element	H	Li	Be	B	C	N	O	F	Na	Mg	Al	Si	P	S	Cl	K	Ca	Sc	Ti	V	Cr	Mn	Fe	Co	Ni	Cu	Zn	Ga	Br	I
VVS	1	3	4	5	6	7	8	9	11	12	13	14	15	16	17	19	20	21	22	23	24	25	26	27	28	29	30	31	35	53
0000	⊕	⊕	⊕						⊕	⊕	⊕					⊕	⊕	⊕	⊕	⊕	⊙	⊕	⊕	⊕	⊕	⊕	⊕	⊕		
0100	○	○	⊕						○	⊕	⊕					○	⊕	⊕	⊕	⊕	⊗				○	○	○	⊕		
0200	⊖	⊖	○	⊕					⊖	○	⊕					⊖	○	⊕	⊕	○	⊗	○	⊕	○	○	○	○	⊕		
0300		⊖	⊖	○	⊕	⊕			⊖	⊖	⊖	○	⊕		⊕	⊖	⊖	⊖	○	⊕	⊗				⊕	⊕	⊕	○		
0400		⊖	⊖	⊖	⊖	○	⊕	⊕	⊖	⊖	⊖	⊖	○	⊕	⊕	⊖	⊖	⊖	○	⊕	⊗		⊗	⊗	○	⊗	⊗	⊗		
0500		⊖	⊖	⊗	○				⊖	⊖	⊖	⊖	○	⊕			⊗	⊖	⊖	○	⊕	⊗	⊗	⊗	⊖	⊗	⊗			
0600		⊖	⊖	⊗					⊖	⊖	⊖	⊖	⊖	○			⊗	⊗	⊖	⊖	○	⊗	⊗	⊖	⊗	⊗				⊕
0700																	⊗	⊗	⊖	⊖	⊗		⊗			⊗				○
0800																	⊗	⊖			⊗		⊗			⊗				
1000	⊙	⊙	⊙						⊙	⊙						⊙	⊙										⊗			
1100																⊙														
1200			⊙	⊙								⊙	⊙																	
1300				⊙	⊙								⊙	⊙				⊙												
1400				⊙	⊗								⊙	⊙																
1500																			⊙											
1600																														
1700																														
2000	⊖	⊖	○						⊖	○	⊕					⊖	○		○		⊗	⊗				○	○			
2100			○	⊕							○	⊕																		
2200				○	⊕	⊕				○	⊕																			
2300				⊖	○	⊕					⊖	○	⊕																	
2400					⊖	⊖	○					⊖	⊖	○	⊕							○							⊕	⊕
2500													⊖	○															○	○
2600																														⊖
3000			○									○														⊙				
3100				⊙	⊙	⊙																								
3200				○	⊙	⊙							⊙	⊙															⊙	⊙
3300													⊙	⊙																
3400																														
3500																														
4000				○									○							○										
4100					⊖	○	⊕							○	⊕															
4200					⊖	⊖	○	⊕					⊖	○	⊕														⊕	⊕
4300						⊖							⊗	⊖	○														○	○
4400						⊖	⊖							⊖	⊖														⊖	⊖
5000					⊙	⊙							⊙	⊙									○							
5100					⊙	⊙	⊙							⊙	⊙														⊙	⊙
5200					⊙										⊙															○
5300																														
6000						○	⊕							○	⊕								○						⊕	⊕
6100						⊖	⊖	○						⊖	○														○	○
6200							⊖								⊖														⊖	⊖
7000					⊙	⊙								⊙	⊙														⊙	⊙
7100							⊙								⊙														⊙	⊙
8000						⊖	⊖							⊖	⊖														⊖	⊖
0010			⊕								⊕							⊕				⊗								
0020		⊖	○	⊕								○	⊕					○	⊕	⊗										
0030			⊕										⊖	○	⊕					○	⊗								⊕	⊕
0040															⊖														⊖	
1010				⊙	⊙																	⊗								
1020					⊙																									
1030														⊙															⊙	⊙
2010				○	⊕	⊕					○																			⊕
2020					○								⊖	○		⊕														⊕
2030																														⊕

a

Table 2. (Continued)

Element	H	Li	Be	B	C	N	O	F	Na	Mg	Al	Si	P	S	Cl	K	Ca	Sc	Ti	V	Cr	Mn	Fe	Co	Ni	Cu	Zn	Ga	Br	I
VVS	1	3	4	5	6	7	8	9	11	12	13	14	15	16	17	19	20	21	22	23	24	25	26	27	28	29	30	31	35	53
3010					⊖	⊙	⊙						⊙	⊙																
3020													⊙	⊙															⊙	⊙
4010					⊖	○	⊕					⊖	○	⊕															Φ	Φ
4020														⊖																
5010						⊙																								
6010																														
0110			○	⊕						○	⊕																			
0210					⊖	○	⊕	⊕				⊖	○	⊕	⊕							⊗	⊗	⊗						
0310					⊖	○						⊖	○	⊕								○	⊗		Φ	⊘				
0410					⊖	⊖	○							⊖				⊗	⊗	○										
0510														⊖																○
0610																							⊗							
1110					⊙	⊙	⊙						⊙	⊙																
1210					⊙	⊙							⊖	⊙																
1310																														
1410																														
1510																														
2110					⊖	○	⊕					⊖	○	⊕															⊕	
2210					⊖		○					⊖	○	⊕															Φ	Φ
2310													⊖	○																
2410														⊖																
3110							⊙							⊙																
3210																														
3310																														
4110					⊖	○						⊖	○																○	○
4210																														⊖
5110																														
0120														○	⊕					○		⊗								
0220													⊖	○									⊗	⊗						Φ
0320																														○
0420								⊖																						
1120																														⊙
1220																														
1320																														
2120													⊖	○															○	○
2220																														⊖
3120																														
0130														○									⊗						○	○
1130																														
0230																														
0001				Φ							Φ																			
0101				⊖	○	⊕	⊕				○	⊕																		
0201				⊖	○							○	⊕							○										
0301				⊖										○									⊗							
0401																														
0501																														
1001					⊙	⊙																								
1101					⊙	⊙																								
1201																														
1301																														
0401																														
2001					⊖	○	⊕						⊖	○																
2101													⊖	○																
2201																														
2301																														

b

charge in this table. However, in real aplications within the program PEGAS, a stable valence state is considered in a freer sense as an "observable" valence state.

This makes it possible, for example, also to suggest valence states $-\overset{\displaystyle |}{\underset{\displaystyle |}{N}}{}^{+}-$, $-\overset{\displaystyle |}{\underset{\displaystyle |}{\overline{O}}}{}^{+}-$

etc. as being stable, (realistic).

The valence state of an atom and its formal expression is a static description for the atom. Let us turn our attention to changes of valence states of atoms in the course of a chemical reaction.

3.2 Conversion of Valence States

In the course of a chemical reaction, the valence states of atoms of reaction sites are converted. It can formally be expressed as [28, 39]

$$v_i + \Delta v = v_f \tag{36}$$

where v_i and v_f is the valence state before (initial) and after (final) the chemical change, respectively, and Δv is an operator, also a four-dimensional vector. It is called a *vector of conversion*. It is easy to see that

$$\Delta v = v_f - v_i \tag{37}$$

Example 3.2   Let us consider the change

$$-C \equiv \overline{N} \longrightarrow \overset{\diagdown}{\underset{\diagup}{C}} = N^{+} \overset{\diagup}{\underset{\diagdown}{}}$$

Vectors of conversion for individual atoms are as follows:

$$\Delta v^C = v_f^C - v_i^C = (0, 2, 1, 0) - (0, 1, 0, 1) = (0, 1, 1, -1),$$

$$\Delta v^N = v_f^N - v_i^N = (0, 2, 1, 0) - (2, 0, 0, 1) = (-2, 2, 1, -1).$$

The vector of conversion models a change which takes place in the course of a chemical reaction. It is known from the theory of reaction mechanisms, and also used in quantum chemical studies of chemical reactions, that chemical reaction can be understoo d as a composition of elementary reactions here called *Elementary Conversions of Valence States*, ECVS. Combinatorially, it is possible to derive ECVS [37, 39, 28], which are also considered in our model. All of them together with their notation and reaction schemes are shown in Table 3.

Table 3. Elementary conversions of valence states of atoms, their notation, reaction scheme (formally expressed for general atoms I and J) and corresponding vectors Δv

No	Notation	Reaction scheme	ΔV for reference I atom
		Heterolytic reactions	
1	$1D_E$	I—⏐J ⟶ I⁺ + J	$(2, -1, 0, 0)$
2	$2D_E$	I⩵J ⟶ Ī—J	$(2, 1, -1, 0)$
3	$3D_E$	I⩵J ⟶ Ī⩵J	$(2, 0, 1, -1)$
4	$1D_N$	I⏐—J ⟶ I + J̄	$(0, -1, 0, 0)$
5	$2D_N$	I⩵J ⟶ I—J̄	$(0, 1, -1, 0)$
6	$3D_N$	I⩵J ⟶ I⩵J̄	$(0, 0, 1, -1)$
7	$1A_E$	Ī⁺ J ⟶ I—J	$(-2, 1, 0, 0)$
8	$2A_E$	Ī—J ⟶ I⩵J	$(-2, -1, 1, 0)$
9	$3A_E$	Ī⩵J ⟶ I⩵J	$(-2, 0, -1, 1)$
10	$1A_N$	I⁺ J̄ ⟶ I—J	$(0, 1, 0, 0)$
11	$2A_N$	I—J̄ ⟶ I⩵J	$(0, -1, 1, 0)$
12	$3A_N$	I⩵J̄ ⟶ I⩵J	$(0, 0, -1, 1)$
		Homolytic reactions	
13	$1D_R$	I⫢J ⟶ I· + ·J	$(1, -1, 0, 0)$
14	$2D_R$	I⩵J ⟶ İ—J̇	$(1, 1, -1, 0)$
15	$3D_R$	I⩵J ⟶ İ⩵J̇	$(1, 0, 1, -1)$
16	$1A_R$	İ + J̇ ⟶ I—J	$(-1, 1, 0, 0)$
17	$2A_R$	İ—J̇ ⟶ I⩵J	$(-1, -1, 1, 0)$
18	$3A_R$	İ⩵J̇ ⟶ I⩵J	$(-1, 0, -1, 1)$
		Redox reactions	
19	$1D_0$	I⁺ J ⟶ I + J̇	$(-1, 0, 0, 0)$
20	$2D_0$	Ī⁺ J ⟶ I + J̄	$(-2, 0, 0, 0)$
21	$1A_r$	I⁺ J̇ ⟶ İ + J	$(1, 0, 0, 0)$
22	$2A_r$	I⁺ J̄ ⟶ Ī + J	$(2, 0, 0, 0)$

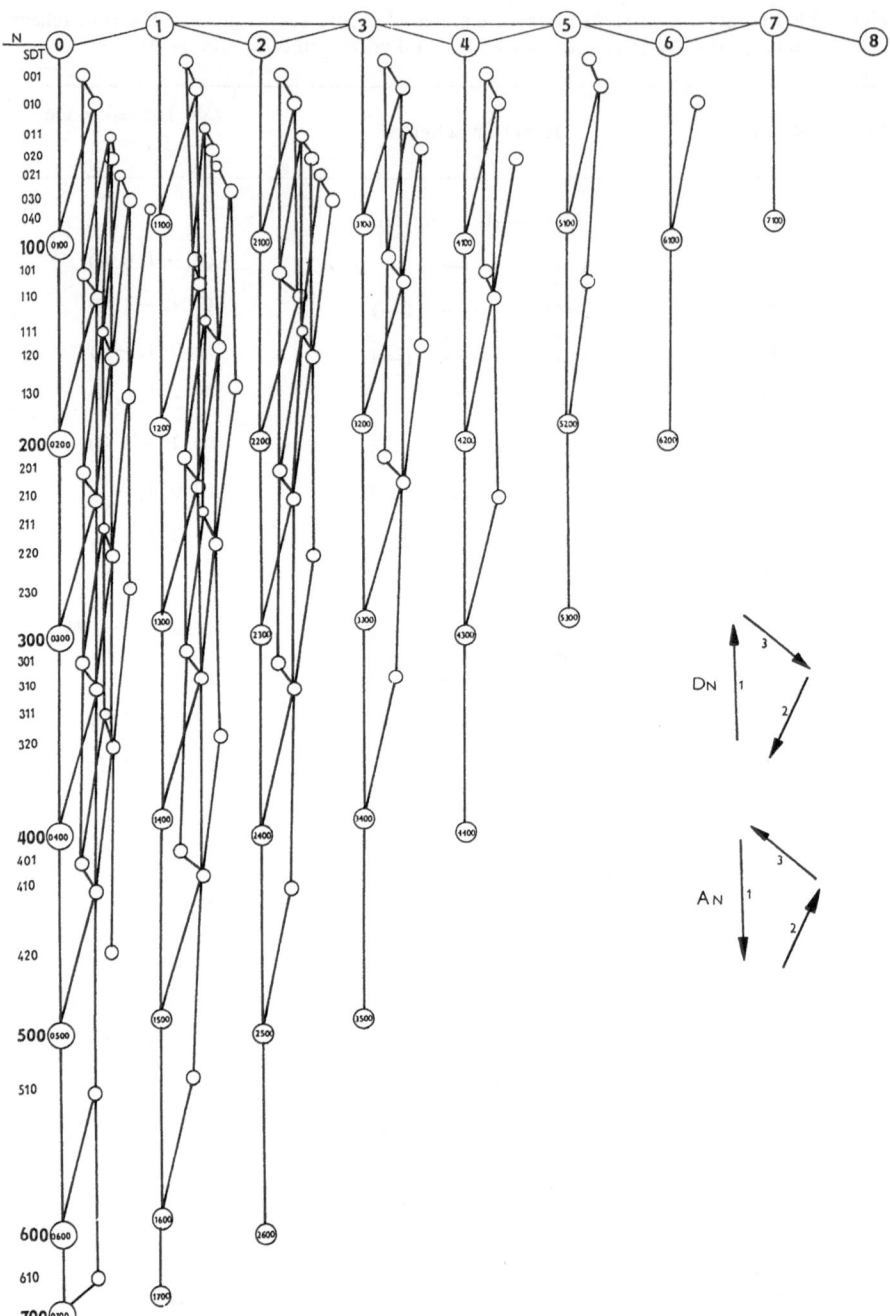

Scheme 1. Subgraph of the graph G_{ECVS} for nucleophilic conversions

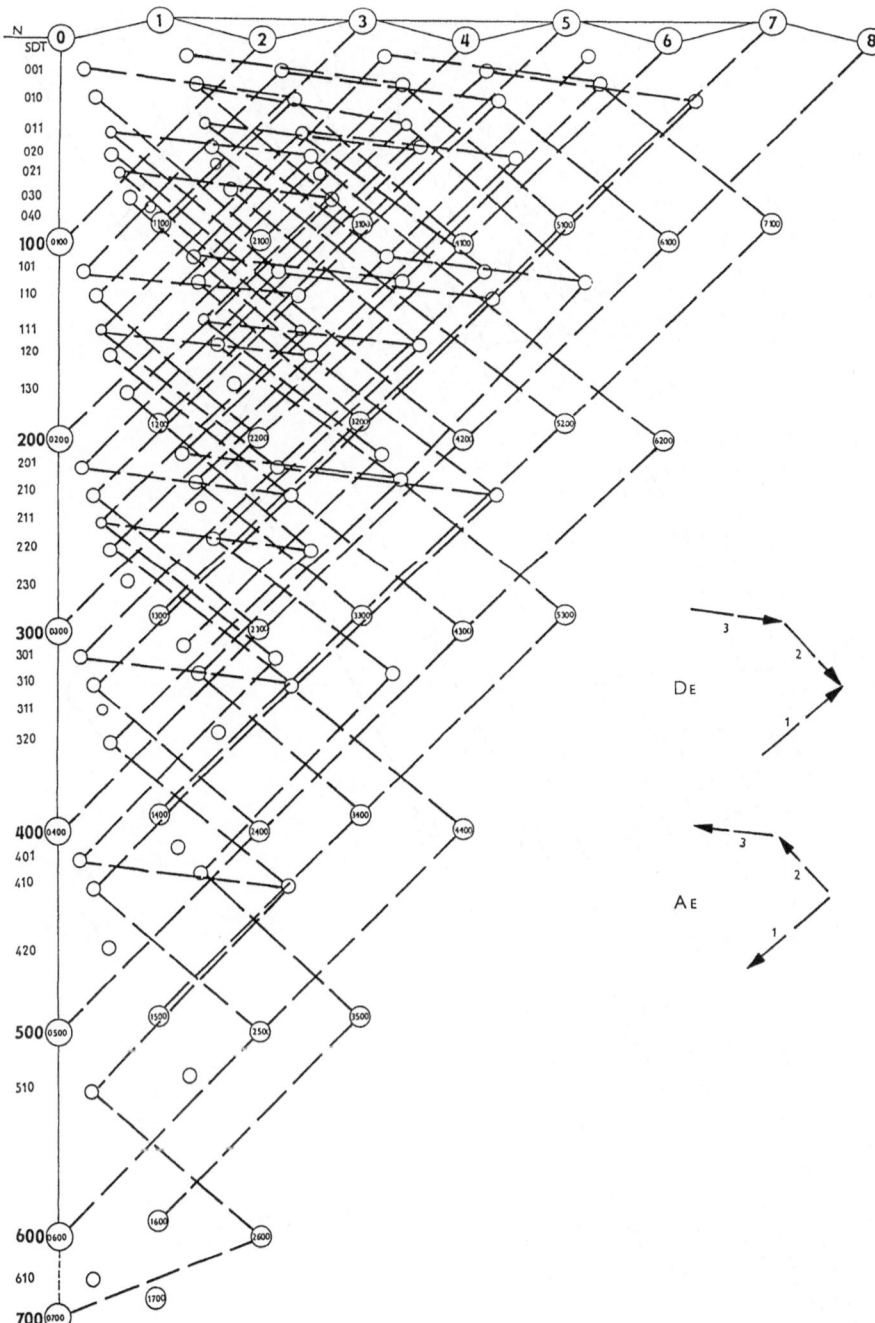

Scheme 2. Subgraph of the graph G_{ECVS} for electrophilic conversions

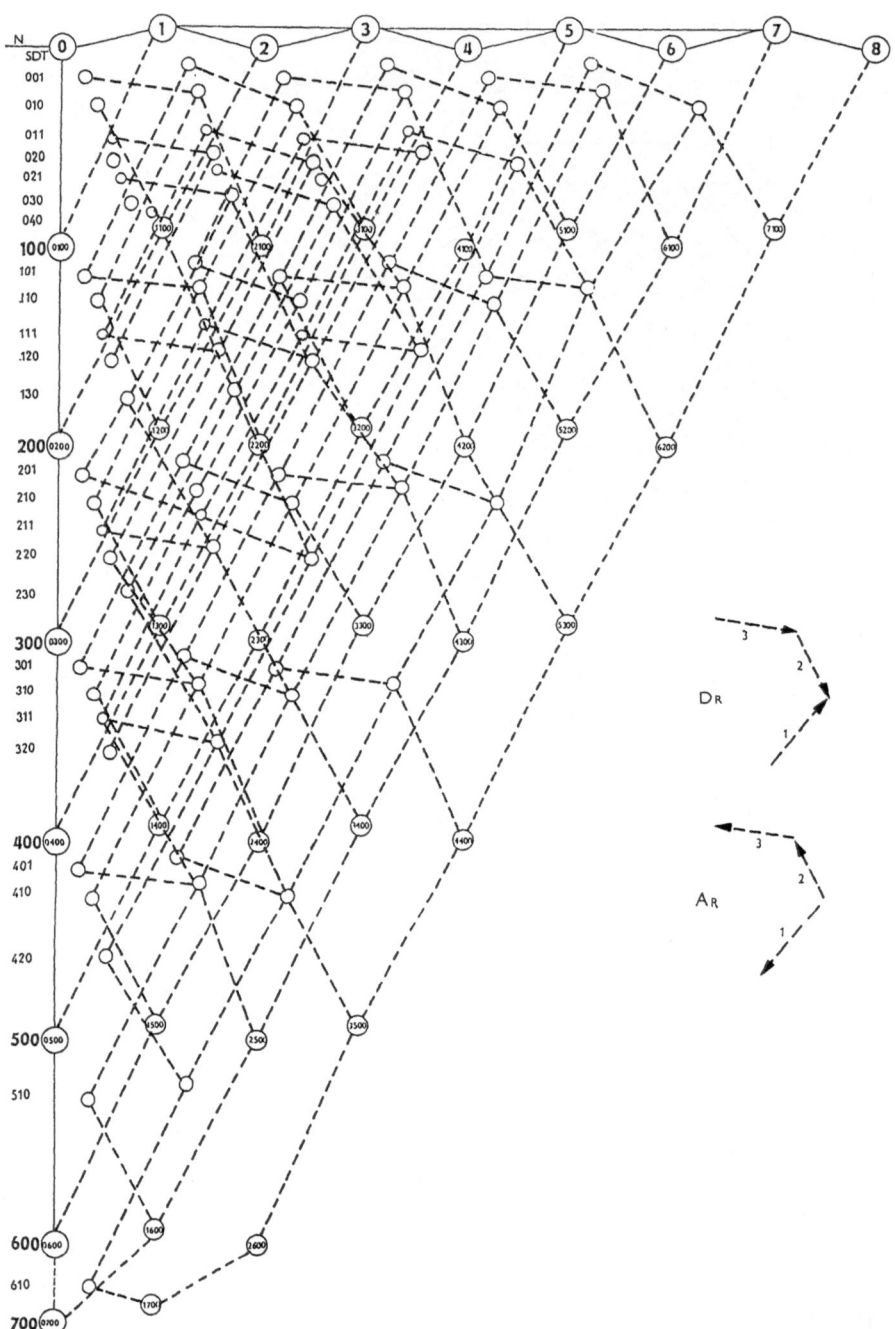

Scheme 3. Subgraph of the graph G_{ECVS} for radical conversions

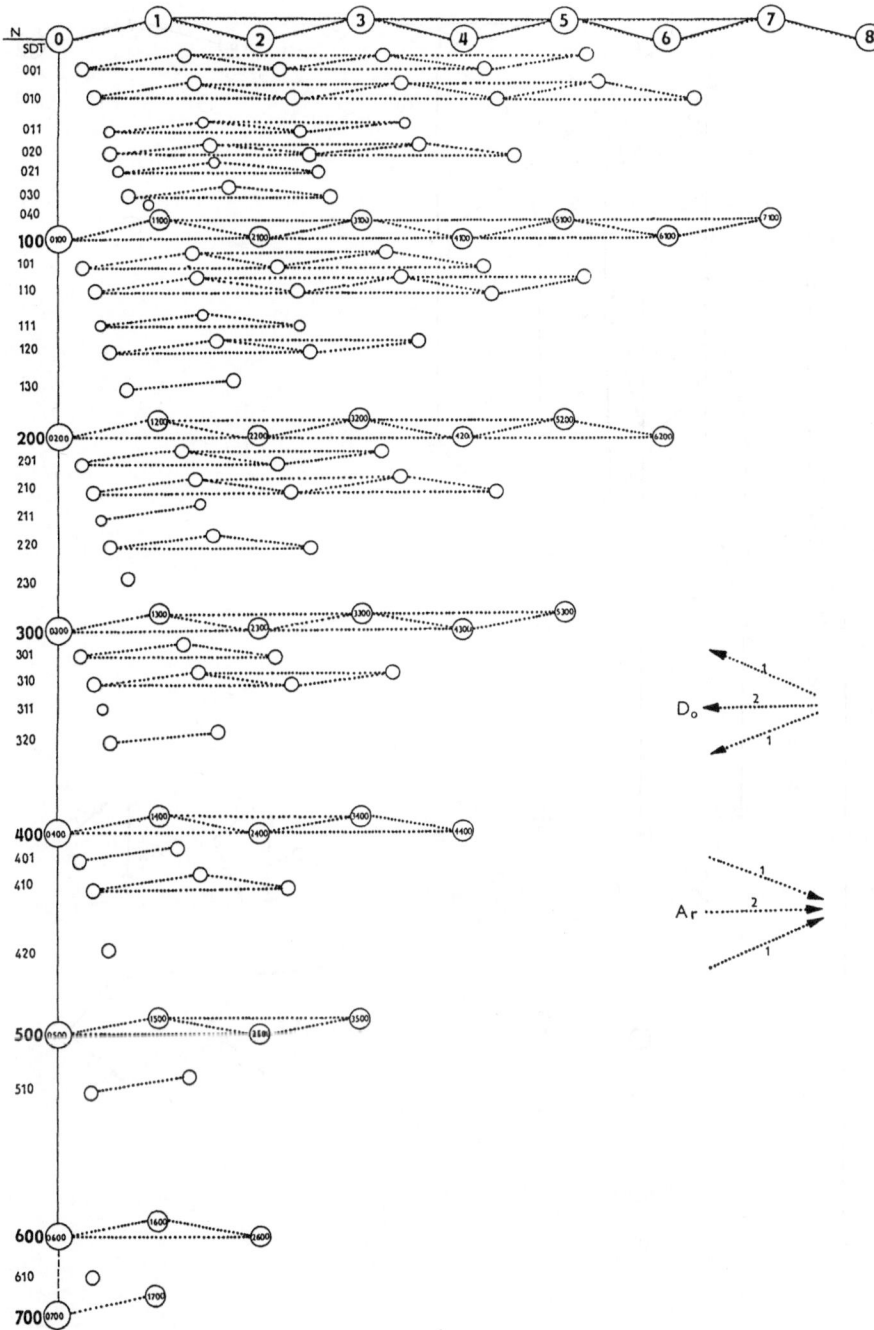

Scheme 4. Subgraph of the graph G_{ECVS} for redox conversions

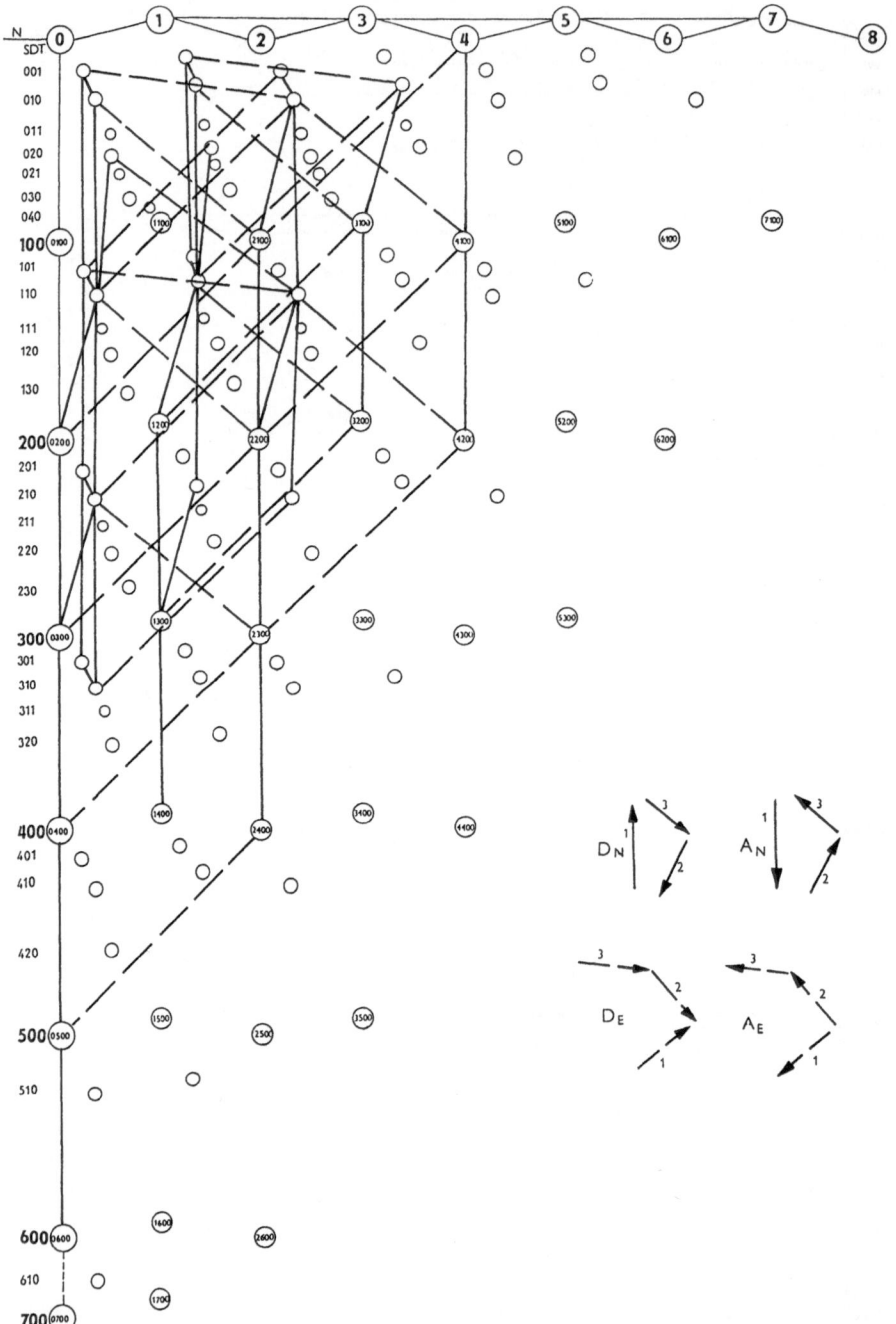

Scheme 5. Subgraph of the graph G_{ECVS} for nucleophilic and electrophilic conversions of carbon

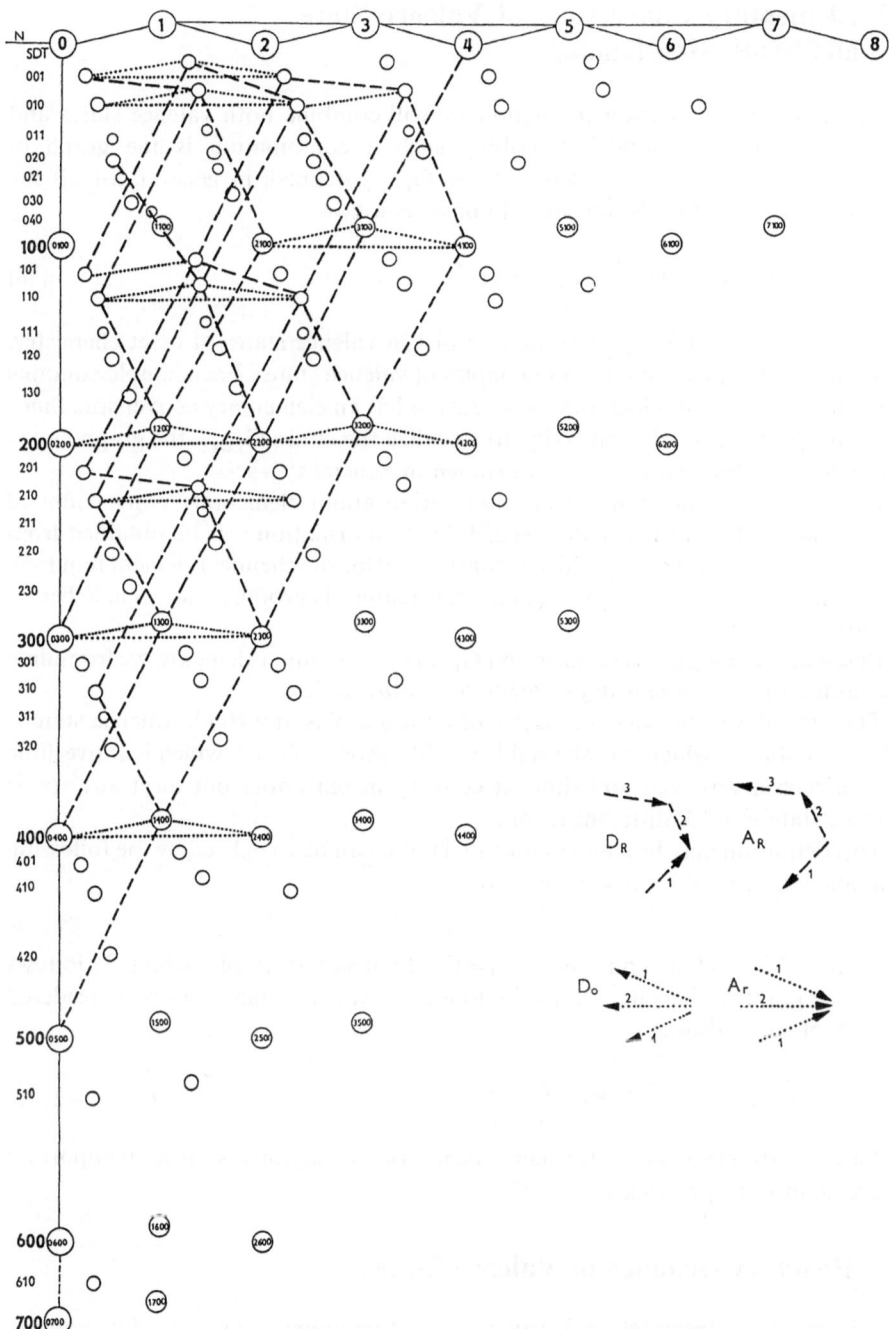

Scheme 6. Subgraph of the graph G_{ECVS} for radical and redox conversions of carbon

153

3.3 Elementary Conversions of Valence States and Synthesis Planning

In order to model a chemical reaction we will combine both valence states and their conversions. A model describing such a combination is the graph of elementary conversions of valence states, G_{ECVS}, defined, in general, for all the valence states and ECVS introduced above as follows:

$$G_{ECVS} = (V_{ECVS}, E_{ECVS}) \tag{38}$$

where the vertex set V_{ECVS} is composed of 136 valence states of octet chemistry. The edge set E_{ECVS} is composed of couples of valence states. Each couple contains two valence states for which the vector Δv models an elementary conversion. Since it is difficult, because of complexity, to visualize the entire G_{ECVS}, the graph G_{ECVS} is divided into four parts which are shown in Schemes 1–4.

For organic synthesis, however, information about elementary conversions of an individual element is usually needed. Such information can be obtained from a suitable subgraph of G_{ECVS} which is constructed for the chemical element required. An example, the subgraph of G_{ECVS} for conversions of carbon, is shown in Schemes 5 and 6.

Based on the subgraphs of the graph G_{ECVS} for individual elements, we formulate the notion of the *synthetical precursor/successor*, SPS.

The SPS of a stable valence state v of an atom X is any stable valence state v' of X (including v) which is in the stable neighborhood of v, i.e. which is convertible to v in such a way that any shortest conversion path does not meet any stable valence state v'' of X different from v.

This definition may be a bit complicated but it can be visualized by the following example based on the carbon subgraph of G_{ECVS}.

Examples 3.3 Let us consider sp^2 ($=\overset{|}{C}-$) valence state of carbon. It follows from the graph in Scheme 5 that the following valence states can be considered as SPS of sp^2 carbon:

$$-\overset{|}{\underset{|}{C}}-, \ -\overline{C}-, \ \equiv C-, \ =C=, \ =\overset{|}{\underset{|}{C}}-, \ \overline{C}=$$

Based on the graph G_{ECVS} we define the notion of reaction distance, an important heuristic in our approach to CAOS.

3.4 Reaction Distance of Valence States

Similarity of molecules [40–44] is an important phenomenon serving, for example, by qualitative and quantitative estimation of chemical reactivity or biological activity of chemical species. Distance is a possible way to measure similarity. In the frame of a mathematical model of the logical structure of chemistry, the notion of chemical distance, CD, has been introduced [13]. The CD is defined as the sum of absolute values of entries of the R-matrix, i.e. as a value corresponding with number of electrons "migrating" in the course of the chemical reaction. The CD is computed from the starting and final state of the system. In order to express also the electron changes during the process we define the notion of reaction distance [20], RD.

Reaction distance of two valence states, v and v', is the length of the shortest path between v and v' in the graph G_{ECVS}. In other words, RD is the minimal number of ECVS which are needed for the conversion $v \rightarrow v'$ or back.

Example 3.4 Let us consider the valence states of carbon: $v: -\overset{\mid}{\underset{\mid}{C}}-$, $v': =\overset{\mid}{C}-$,

$v'': \equiv C-$, $v''': =\overline{C}$. It is seen from the graph in Schemes 5 and 6 that:

$$RD(v, v') = 2, \qquad RD(v, v'') = 4, \qquad RD(v', v'') = 2.$$

$$RD(v, v''') = 4, \qquad RD(v', v''') = 2, \qquad RD(v'', v''') = 2.$$

It is considered from Schemes 5 and 6 that the SPS of a valence state v could also be defined, in most cases, as a set of stable valence states v' for which $RD(v, v') \leq 2$.

An analytical expression for the RD of two valence states was found. It is a special case of the algorithm for the RD of atomic couples (atomic vectors) computation [45].

Note that above formal valence states apparatus can be used, also without any assistance of computer, in new valence states prediction or in mechanistical studies of chemical reactions (e.g. looking for the shortest reaction path). Some possible applications are discussed in [37, 39, 46, 47] in a more detailed manner.

3.5 Synthons

In this section, the above valence states concept will be extended for molecular systems which are called synthons here.

Initially, the notion of the synthon was, in the frame of formalization of synthetical chemistry, introduced by Corey [23]. Based on the Corey approach, and on the Dungundji-Ugi formalization of chemistry [13], the synthon model of constitutional chemistry was introduced [18, 19, 21, 22, 24, 16]. In general, it can be understood as a generalized concept of valence states of atoms and their combinations.

The synthon $S(A)$ over an atomic set A is defined as one or several molecules (or their parts) composed of atoms from the set A. In contrast to the definition of the *Ensemble of Molecules*, $EM(A)$ [13], $S(A)$ is more general because it may also involve free valences, i.e. bonds which do not connect two atoms but only start from an atom. The set of all synthons constructed over the atomic set A is called the *Family of Isomeric Synthons* and is denoted $FIS(A)$. Implicitly, it means that two synthons are called isomeric if each of them is constructed via the same atomic set A.

Formally, a synthon $S(A)$ is described by the S-matrix. Off-diagonal entry a_{ij} of the S-matrix is defined as the multiplicity of the bond between the i-th and j-th atom (non-bonded atoms are bonded by zero-multiplicity bond). Diagonal entry a_{ii} is defined as the four-dimensional vector of valence state of the i-th atom in $S(A)$.

Example 3.5 Let us consider the set $A = \{H, C, N, O\}$. The following structures are synthons from FIS(A).

$$
\begin{array}{ccc}
\text{H}-\text{C}-\overline{\text{N}}{\Large{\diagup}} & \text{H}-\overline{\text{N}}= + -\overset{|}{\underset{|}{\text{C}}}-\overline{\underline{\text{O}}}= & -\overset{|}{\underset{|}{\text{C}}}- + =\overline{\text{N}}-\overline{\underline{\text{O}}}-\text{H} \\
\quad\overset{\|}{|\text{O}|} & & \\
S_1(A) & S_2(A) & S_3(A)
\end{array}
$$

The S-matrix M of synthon $S_1(A)$ may be written as follows

$$
M = \begin{array}{c} \\ \text{H} \\ \text{C} \\ \text{N} \\ \text{O} \end{array}
\begin{array}{cccc}
\text{H} & \text{C} & \text{N} & \text{O} \\
\left[\begin{array}{cccc}
(0, 1, 0, 0) & 1 & 0 & 0 \\
1 & (0, 2, 1, 0) & 1 & 2 \\
0 & 1 & (2, 3, 0, 0) & 0 \\
0 & 2 & 0 & (4, 0, 1, 0)
\end{array}\right]
\end{array}
$$

3.6 Synthon Reactions

A synthon reaction is understood as a process when a starting synthon $S(A)$ is changed to a final synthon $S'(A)$. It is seen that $S(A)$ and $S'(A)$ are members of the same FIS(A), i.e. they have to be isomeric. The synthon reaction is modelled by matrix equation

$$M + R = M' \tag{39}$$

where M and M' is the S-matrix of the starting and final synthon, respectively, and R is the so-called SR-matrix [18, 21]. The additive operation $+$ is understood as an addition of vectors and individual elements for diagonal and off-diagonal entries, respectively.

From Eq. (39) we have simply

$$R = M' - M \tag{40}$$

Example 3.6 Let us consider a reaction change $S_1(A) \Rightarrow S_2(A)$ of the synthon from example 3.5. Its SR-matrix R is expressed:

$$
R = \begin{array}{c} \\ H \\ C \\ N \\ O \end{array}
\begin{array}{cccc} H & C & N & O \\ \left[\begin{array}{cccc} (0,1,0,0) & 0 & 1 & 0 \\ 0 & (0,4,0,0) & 0 & 1 \\ 1 & 0 & (2,1,1,0) & 0 \\ 0 & 1 & 0 & (4,1,1,0) \end{array} \right] \end{array}
$$

$$
- \begin{array}{c} \\ H \\ C \\ N \\ O \end{array}
\begin{array}{cccc} H & C & N & O \\ \left[\begin{array}{cccc} (0,1,0,0) & 1 & 0 & 0 \\ 1 & (0,2,1,0) & 1 & 2 \\ 0 & 1 & (2,3,0,0) & 0 \\ 0 & 2 & 0 & (4,0,1,0) \end{array} \right] \end{array}
$$

$$
= \begin{array}{c} \\ H \\ C \\ N \\ O \end{array}
\begin{array}{cccc} H & C & N & O \\ \left[\begin{array}{cccc} (0,0,0,0) & -1 & 1 & 0 \\ -1 & (0,2,-1,0) & -1 & -1 \\ 1 & -1 & (0,-2,1,0) & 0 \\ 0 & -1 & 0 & (0,1,0,0) \end{array} \right] \end{array}.
$$

Each synthon reaction can be seen as a composition of elementary synthon reactions. It was shown recently that this composition is not unambiguous [21]. Elementary synthon reactions are formally expressed by elementary matrix operators which are summarized in Table 4. Only a symbolic description is given, for the matrix form see [16, 18, 21].

Table 4. Elementary matrix operators and elementary synthon reactions modeled by them

Operator	Process to be modeled	Operator	Process to be modeled
α_1^{ij}	$I - J \rightarrow \bar{I} + J$	γ_2^{ij}	$I = J \rightarrow I^\bullet - J^\bullet$
α_2^{ij}	$I = J \rightarrow \bar{I} - J$	γ_3^{ij}	$I \equiv J \rightarrow I^\bullet = J^\bullet$
α_3^{ij}	$I \equiv J \rightarrow \bar{I} = J$	γ_1^{ii}	$I - \rightarrow I^\bullet$
α_1^{ii}	$I - \rightarrow \bar{I}$	γ_2^{ii}	$I = \rightarrow I^\bullet -$
α_2^{ii}	$I = \rightarrow \bar{I} -$	γ_3^{ii}	$I \equiv \rightarrow I^\bullet =$
α_3^{ii}	$I \equiv \rightarrow \bar{I} =$	δ_1^{ij}	$I^\bullet \mid J \rightarrow I + J^\bullet$
β_1^{ii}	$I - \rightarrow I$	δ_2^{ij}	$\bar{I} + J \rightarrow I + J$
β_2^{ii}	$I = \rightarrow I -$	δ_1^{ii}	$I^\bullet \rightarrow I$
β_3^{ii}	$I \equiv \rightarrow \bar{I} =$	δ_2^{ii}	$\bar{I} \rightarrow 1$
γ_1^{ij}	$I - J \rightarrow I^\bullet + J^\bullet$		

Note that the elementary synthon reactions are formally the same as elementary conversions of valence states. The only principal difference is that the elementary synthon reactions differentiate between electronic processes inside the synthon, and processes between the synthon and its neighborhood. For some mathematical features of S- and SR-matrices see [21].

3.7 The graph of FIS(A) and SPS of a Synthon from FIS(A)

A global picture of electronic changes between synthons of FIS(A) is given by the graph of FIS(A), $G_{FIS(A)}$. It is defined analogically to the graph of elementary conversions of valence states, G_{ECVS} (see Sect. 3.3) as an ordered couple

$$G_{FIS(A)} = (V, E), \tag{41}$$

where $V = \{S_i(A) \subset FIS(A)\}$ is the vertex set, and the edge set E is composed of such couples of synthons from FIS(A) which can be mutually converted by single elementary synthon reaction. An example of a part of the $G_{FIS(A)}$ for $A = \{C, C, P, O\}$ is shown in Scheme 7.

All synthons $S(A)$ from FIS(A) can be divided into two groups and classified as stable and unstable. Following the idea of stable valence states (Sect. 3.1) we say that a synthon $S(A)$ is stable if each of the atoms of $S(A)$ is in a stable valence state in $S(A)$. In all other cases the synthon is called unstable. However, the notion of the stable synthon is relative, and it does not mean that every stable synthon is a "chemically stable compound". This notion just makes the model more flexible. For example, when working in carbene-chemistry we will consider carbene valence states $-\overline{C}-, \overline{C}=$ of carbon as stable, otherwise, seen from an experimental chemist's point of view, they are considered as unstable. When considering only valence states without formal charge as stable then only the synthons 1, 7, and 11 in Scheme 7 are stable.

Based on the $G_{FIS(A)}$ graph, the notion of the SPS of a synthon $S(A)$ could be defined in the same way as that of the SPS of a valence state. However, in this case it would lead, in most cases, to a combinatorial explosion, i.e. the number of SPS of $S(A)$ would be so large that any interpretation would not be possible. In order to make the formulation of SPS practically usable, we have to consider certain limitations. Trying to mimic the way of thinking of the organic chemist, we will suppose that all the parts of the synthon are not of the same importance when inspecting chemical behavior, for example reactivity, and also taking into account various conditions. So, the entire synthon is divided into three parts — reaction center (a subset X of the atomic set A), "intact" or "dummy" part (a subset $\bar{X}$ of the set A) and the rest of the synthon. Excluding the "dummy" part, the so-called active part $S(Y)$ is obtained ($Y = A \setminus \bar{X}$). The exact definition of the SPS, really used in the program PEGAS, is discussed in a more detailed manner in Sect. 4, and exactly introduced in [22]. Here we would like to describe just general ideas.

First, we define a stable neighborhood of $S(A)$ as all synthons $S'(A)$ from FIS(A) for which:

- $S'(A)$ is a stable synthon,
- the "dummy" part of $S(A)$ and $S'(A)$ is the same, i.e. the multiplicities of bonds between atoms of X and valence states of atoms of X are preserved,
- any shortest conversion path $S'(A) \Rightarrow S(A)$ in $G_{FIS(A)}$ does not meet any stable synthon $S''(A)$ ($S''(A) \neq S'(A)$, $S''(A) \neq S(A)$).

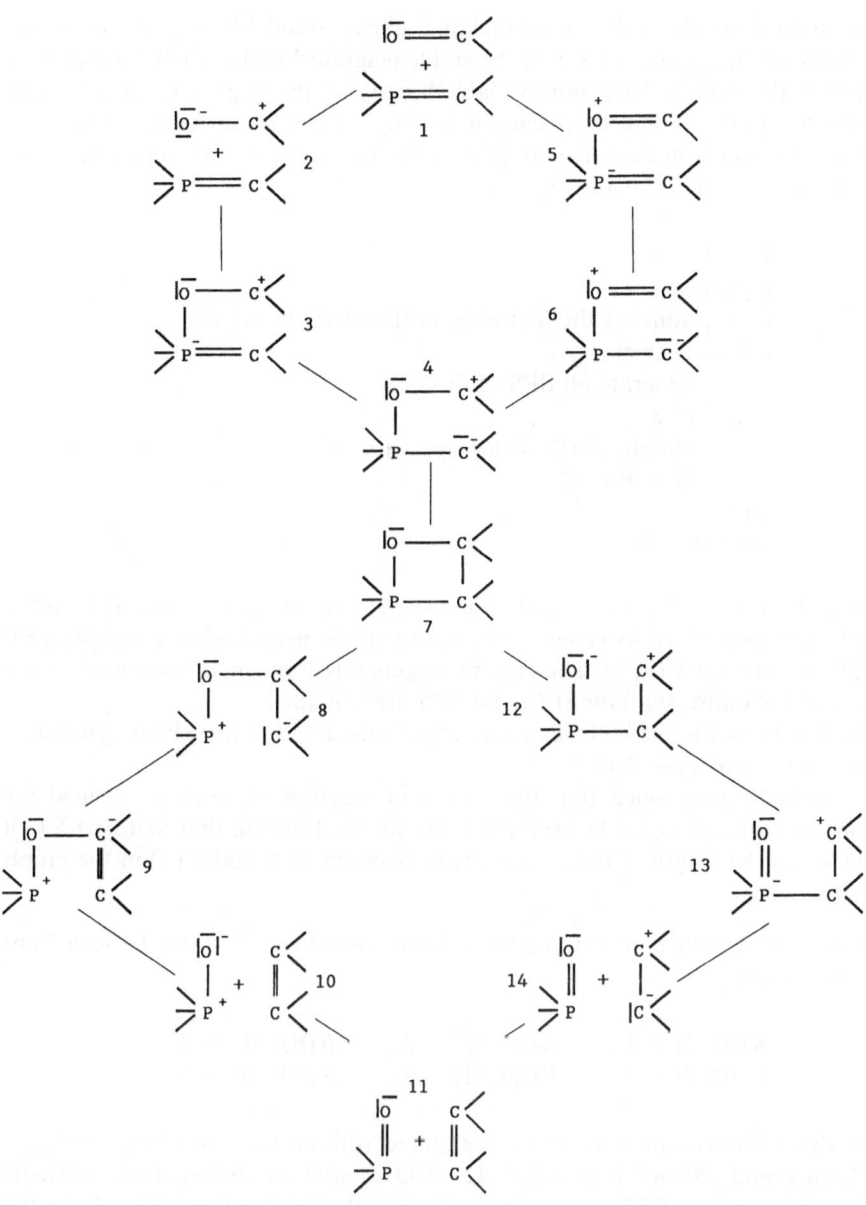

Scheme 7. A part of the graph $G_{FIS(A)}$ for $A = \{C, C, P, O\}$

The SPS of the synthon $S(A)$ is then each synthon $S'(A)$ of the neighborhood for which the electronic change $S(A) \Rightarrow S'(A)$ is "smooth", i.e. it is not separated into two or more mutually isolated electronic processes, and the difference between multiplicity of corresponding bonds in $S(A)$ and $S'(A)$ is less than two. The formulation of the SPS is also slightly dependent on the set X (see section 4).

Eva Hladka et al.

The notion of the SPS as formulated above implicitly expects complete knowledge of $G_{FIS(A)}$ or, at least, of the stable neighborhood of $S(A)$. However, in most cases this knowledge is not available because of the large size. Therefore, an alternative approach has been chosen leading to the same results. The idea, discussed in details in Sect. 4 or in [22], is composed of several steps which can be expressed by the following algorithm.

```
B := X
repeat
C := {atoms of the first neighborhood of the set B}
if B = X then
        generate all SPS of S(X)
        else
        stabilize S(B) with respect to S(C)
        B = B ∪ C
end
until B = A
```

Since the set X, as seen in Sect. 5, is usually one or two atomic, it is not a complicated procedure to generate the entire stable neighborhood and then all the SPS on the set $FIS(X)$. The structures generated are then "stabilized" with respect to the entire synthon $S(A)$, and SPS are obtained.

The results are illustrated by Scheme 7. Synthons 1 and 11 are SPS of synthon 7. For more examples see Sect. 5.

It should be mentioned that the notion of reaction distance, as defined for valence states of atoms, can be used also for synthons. It means that $RD(S(A), S'(A))$ is defined as the length of the shortest path between $S(A)$ and $S'(A)$ in the graph $G_{FIS(A)}$.

Example 3.7 Let us consider synthons shown in Scheme 7. It can be seen from the graph that:

$$RD(1, 2) = 1, \qquad RD(1, 3) = 2, \qquad RD(1, 4) = 3,$$
$$RD(4, 7) = 1, \qquad RD(4, 11) = 5, \qquad RD(1, 11) = 8.$$

An algorithm by which the RD is computed without any knowledge of $G_{FIS(A)}$ has been found [20] (cf. Sect. 4.2.2). The RD is used as an important heuristic during the process of SPS generation. It also plays a fundamental role in the program for reaction network generation which is now being developed.

4 Computer Realization

The theory of the synthon model of organic chemistry, presented in previous section, found its realization in the computer program PEGAS (**P**rogram for **E**lective **G**eneration of **A**ccessible **S**ynthons) [48]. This section is devoted to the decription of the PC-based PEGAS implementation as well as to the description of main algorithms used in the implementation.

The package PEGAS is an interactive program that has been implemented for IBM-compatible personal computers using the MS-DOS operating system. It is an *expert system* in the sense that it assumes the interaction with a skilled user-expert-who exerts his/her professional skill to choose acceptable reactions and reaction paths. PEGAS is able to generate proposals of chemical reactions in the forward direction (from the given educts to products) as well as in the retro-direction (from the target compound to precursors). PEGAS generates the (retro)synthetic tree with a reasonable number of branches up to any desired level. Examples of the use of the program package PEGAS are given in Sect. 5.

4.1 Program Structure

The program package PEGAS consists of several relatively independent modules:
— The user-friendly main menu driver, which allows the graphic input of educt and/or product structures, management of the (retro)synthetic tree (traversal through and display of nodes), display of structures generated by the SPS generator, selection of acceptable reactions and reaction paths, and similar interactive tasks.
— The SPS generator MAPOS.
— The special filters used to filter-out structures with forbidden substructures, or structures created through forbidden reaction processes.
— The editor of valence states of atoms used in the SPS generator.
This arrangement saves computer main memory allowing the generation of synthon precursors/successors of large structures. All information transferred between modules is stored in external files.

4.1.1 Menu Driver

All interactive activity is controlled by the main-menu driver, which also runs other programs as requested. The input/output options of this menu driver allows one:
— to draw a new structure, and store it in the file,
— to retrieve a structure from a file, to edit it, and
— to display structures from the SPS generator, to modify, and to save them.

All structures are drawn on the rectangular grid with the step size specified by the user. Although the synthon-model algorithms work only with the topology of chemical compounds (i.e. no three-dimensional structure information is needed and/or utilized), the two-dimensional structure of the drawn compound is also stored, and it is used for display of compounds generated by the SPS generator. The drawing program uses the stored two-dimensional structure of the starting compound(s) for two main reasons:
1. It is generally very difficult to restore the two-dimensional structure (the "picture") of any non-trivial chemical compound having only the topology of

this compound available, and the generation of the two-dimensional structure is rather time-consuming task [49]. Use of the stored template simplifies substantially the drawing task.

2. The user of the program package must be able to recognize quickly what happened with the starting chemical compound during the suggested transformation. In general, when fragmentation occurs, it is almost impossible to create a new two-dimensional structure similar enough to the two-dimensional structure of the starting compound what is essential for easy identification of common parts of both structures, without the knowledge about the two-dimensional structure of the starting compound.

During the inspection of the structures generated by the SPS generator, the user could edit all structures which do not fulfill his/her esthetic requirements.

The user may also use a library of predefined generic structures — "templates" — for drawing of complex (e.g. polycyclic) compounds, and even to add new templates into this library.

The input/output driver is also used for the (retro)synthetic tree inspection. The user can walk through the tree generated, inspect the selected nodes, edit them, and submit them for further processing by the SPS generator, thus creating new branches and levels of the (retro)synthetic tree. Each node of the (retrosynthetic trees stores not only the two-dimensional structure of the chemical compound, but also the distance from its "father" (i.e. starting structure) measured by the reaction as well as chemical distances, and also its fragmentation level, measured by its order.

The valence states of all atoms of any structure drawn or modified by the user are checked, charged atoms as well as atoms with odd number of electrons are colored, and valence states not found in the table of permitted valence states for a chemical element are reported. Structures may be drawn without the hydrogen atoms, these may be added automatically by the system (although only for carbon, nitrogen, phosphorus and sulfhur in their most usual valence states with a closed valence shell). The display of hydrogen atoms may also be switched on or off.

During the preparation of structures for the SPS generator, the user must select atoms of the *reaction center* (active atom(s)), and could select the *passive* (unchangeable) part of the structure. It is possible to select several distinct reaction centres on the same structure. When the selection of the active atoms not performed, the user is warned and the program does not allow the storing of the structure until at least one-atom reaction centre is selected. This behavior ensures that no structure without a defined reaction centre is eventually submitted to the SPS generator.

The user-friendliness is enhanced by the context-sensitive help, available in all modes of processing. The help window may be permanently present on the screen (reducing the part of the screen used for draw/display of structures) or it may be called in only when actually needed. When the help window is opened, its contents are updated automatically with the selection of menu and submenu items.

4.1.2 SPS Generator

The SPS generator is the executive heart of the whole program package PEGAS. The generator used is, in fact, a MAPOS program [50], which generates the synthetic successors/precursors of the starting structure. The generation is performed according to the synthon model of organic chemistry, described in the preceeding sections, the actual formulation of algorithms used is presented in the next part of this section.

The program MAPOS runs in batch mode. It is started whenever the user selects the SPS generation in the main menu, it generates all the synthetic precursors/successors of a given compound, stores them in an external file, and returns control to the main-menu driver.

According to the theory, MAPOS works in two successive steps:
1. Generation of all permitted transformations of the reaction center.
2. Stabilization of each transformed reaction center.

The set of all transformed reaction centres is generated by a straightforward implementation of the usual combinatorial algorithm (cf. Sect. 4.2.3), and the transformed reaction centers are checked to fulfill several restrictive conditions which allow one to suppress the possibility of combinatorial explosion at the very beginning of the SPS generation. The generation of transformed reaction centers is very fast (it usually takes no more then 10–15% of the overall execution time when the reaction center has no more then 4 atoms in a molecule with more then 10–15 atoms). Only non-isomorphic transformed reaction centers are processed by the second step.

Each transformed reaction center generated in the first step of the program MAPOS is stabilized using the algorithm STAB (cf. Sect. 4.2.4). As the STAB algorithm does not produce only non-isomorphic structures, all the generated structures are checked for isomorphism with the ones generated previously, and only non-isomorphic structures form the output of the SPS generator.

Processing capabilities of the MAPOS program are limited only by the computer memory available. MAPOS allocates the whole memory of your computer (in the DOS available area, of course) for the heap, and the allocation of the heap memory is managed by the proprietary allocation algorithm. Information about all the synthons generated during the execution is stored on the heap, and this memory is released as soon as possible. Memory available on the standard PC (with 640 kB of main memory, and only with operating system (version 3.30 or similar) and mouse driver loaded) is sufficient for allocation of several hundreds (more then 500) of 20-atom synthons. As the memory requirements depend on many factors, not only the size of the starting synthon (measured by the number of its atoms) determines the limits of usability of the program. The size of the reaction center, the size of the "passive" part of the compound(s), and also the number and kind of valence states permitted for individual elements present in the compound(s) are the main secondary factors with strong influence on the memory requirements. When processing large molecules (with multiatom reaction centers), the computation time on a PC may be a more limiting factor then the memory available (although generation of 300 15-atom synthons may take no more then 20 minutes on a 12 MHz PC-AT).

4.1.3 Valence States Editor

The valence states of atoms represent building bricks of the whole synthon model of organic chemistry. All chemical reactions are seen as valence electron reorganizations modifying valence states of individual atoms in reacting compounds. The number and kind of valence states permitted for particular chemical elements have a strong influence on the number and kind of synthon precursors/successors generated by the SPS generator MAPOS. The program package PEGAS maintains the table of valence states of all atoms from the periodic table up to and including Neptunium (although entries for some atoms are empty). The valence states editor is a mouse-driven program used for modification of this valence states table.

All valence states of chosen chemical elements are displayed on the display, and color is used to distinguish between possible kinds of valence states — stable, permitted, and forbidden. The user may change the classification of any valence state with the mouse. It is very easy to restrict the number of generated SPS of any chemical compound by reducing the number of stable valence states for the chemical elements from which the compound is composed. Note that only the specification of stable valence states have any real influence on the output of the whole package as may be seen from the definition of the SPS (The SPS is formed by *stable* synthons). On the other hand, selection of unusual valence states as stable may be used during the reaction mechanism analysis, when the generation of intermediates is required. Studies in carbene chemistry may serve as a good example when the change of the default valence states of carbon is requiered.

4.1.4 Filters

Due to the general nature of the underlying model, MAPOS is unable to generate just transformations of some pre-specified kind. In cases when only a special kind of transformation is required, structures generated by MAPOS must be filtered out, and only those fulfilling specified requirements are kept. There are currently 4 different filters:

1. NO-CHARGE means that all structures with at least one atom with a formal charge (i.e. the "charged" structures) are discarded.
2. CHARGE is the complement to the above mentioned filter, i.e. only structures with at least one atom with a formal charge are retained.
3. NON-REDOX means that all structures generated through some redox process, are discarded.
4. CYCLIZATIONS means that not only structures generated through some redox process are discarded as above, but new structures may be generated by this filter as well. This is the only filter which may, in some cases, enlarge the number of SPSs generated from the starting compound. MAPOS is unable to generate new bonds between two atoms unless both these two atoms are members of the reaction center. Although in principe any atom may become a part of a reaction centre, it is sometimes useful (and more efficient) to allow a "collapse" of virtual atoms — if the generated synthon contains two virtual atoms connected to two different atoms by a bond of the same order, it is possible to discard both virtual atoms and create a new bond between the

"real" atoms. A collapse of virtual atoms mimics the rearrangements and cyclizations (as the name of this filter suggests). These transformations can not be modeled by MAPOS in a single step (they require two successive steps).

4.2 Algorithms

In previous sections almost all the basic algorithms of the synthon model of organic chemistry were briefly sketched. In this subsection we will present a more rigorous and detailed definition of the key algorithms, as the exact definition is needed for deep understanding of the possibilities, and limitations, of the synthon model.

Only algorithms internal to the synthon model are presented, although many others are needed for the implementation of the SPS generator. One of the most important algorithms not presented here is the algorithm for the canonical indexation of synthons (enabling the removal of isomeric structures). In the program package PEGAS, a slightly modified version of Ugi's indexing algorithm [51] is used. The existence of virtual atoms in our model forced the modification, but it is not an important one, and the behavior for structures without virtual atoms is identical to the original Ugi's algorithm. The modification used requires strict matching of virtual atoms, so only structures with an equal number and kind of atoms and with an equal number of virtual atoms may be isomorphic.

4.2.1 SR-Graph Construction

The problem is as follows [19]:
Given two isomeric synthons $S(A)$ and $S'(A)$ construct all synthon-reaction graphs (SR-graphs) for the transformation

$$S(A) \Rightarrow S'(A).$$

Due to the existence of the virtual atoms without unique mutual mapping, there may be more than one SR-graph for the specified transformation. The unique internal part of the SR-graph is a graph G_R^I defined as follows:
Let $S(A) = G^I \cup G^E$, and $G^I = (V^I, E^I, L, \varphi, \vartheta)$, $S'(A) = G'^I \cup G'^E$, and $G'^I = (V^I, E'^I, L', \varphi', \vartheta)$ where V^I is the set of non-virtual vertices of $S(A)$, and E^I is the set of edges which are not incident with virtual atoms. Then the *internal part* of the SR-graph is

$$G_R^I = (V_R, E_R^I, L_R, \psi^I, \{+1, -1\})$$

where $E_R^I = E^I - E'^I$, $L_R = L - L'$, $\psi^I: E_R \cup L_R \mapsto \{+1, -1\}$,

$$\psi^I(e) = \begin{cases} -1 & \text{for} \quad e \in E^I \\ +1 & \text{for} \quad e \in E'^I \end{cases}, \qquad \psi^I(l) = \begin{cases} -1 & \text{for} \quad l \in L^I \\ +1 & \text{for} \quad l \in L'^I \end{cases},$$

and the vertex set $V_R \subset V$ is composed of all vertices incident with an edge or loop from the sets E_R^I and L_R, respectively.

For the construction of the (generally non-unique) external part of the SR-graph we must introduce the following notions.

Let $v = (0, v_1, v_2, v_3)$ be a vector of *external* valence states of atom $A_i \in S(A)$, i.e., v_1 is the number of single bonds from the atom A_i to virtual atoms, v_2 is the number of double bonds to virtual atoms, and v_3 is the number of triple bonds to virtual atoms. Let $u = (0, u_1, u_2, u_3)$ be the vector of external valence states of the same atom A_i but in the synthon $S'(A)$. Put

$$w = (w_1, w_2, w_3) = (u_1 - v_1, u_2 - v_2, u_3 - v_3)$$

The vector v can be expressed in the x, y, z basis of the $E^{(3)}$ space, where $x = (-1, 0, 0)$, $y = (1, -1, 0)$, and $z = (0, 1, -1)$ model extinction of a single, a double, and a triple bond, respectively

$$w = a(-1, 0, 0) + b(1, -1, 0) + c(0, 1, -1),$$

i.e.

$$a = -w_1 - w_2 - w_3, \qquad b = -w_2 - w_3, \qquad c = -w_3.$$

The computation of the external part of the SR-graph is based on the minimization of this graph on individual atoms. The minimal number of created external bonds N_+, and the minimal number of vanished external bonds N_- at atom A_i during the $S(A) \rightarrow S'(A)$ transformation is

$$N_+ = F(-a) + F(-b) + F(-c)$$

$$N_- = F(a) + F(b) + F(c)$$

where

$$F(x) = \begin{cases} x & \text{for } x \geq 0 \\ 0 & \text{for } x > 0 \end{cases}.$$

There exist only 12 different combinations $s_1, s_2, \ldots, s_{12}$ of the a, b, c coordinates for an atom A_i and any fixed virtual vertex [19]:

	s_1	s_2	s_3	s_4	s_5	s_6	s_7	s_8	s_9	s_{10}	s_{11}	s_{12}
a	1	0	0	1	0	1	-1	0	0	-1	0	-1
b	0	1	0	1	1	1	0	-1	0	-1	-1	-1
c	0	0	1	0	1	1	0	0	-1	0	-1	-1

The vector (a, b, c) may be expressed by the means of vectors $s_1, s_2, ..., s_{12}$

$$(a, b, c) = \sum_{i=1}^{12} t_i s_i$$

where t_i are parameters. The N_+ and N_- may be described in means of these parameters

$$N_- = t_1 + t_2 + t_3 + 2t_4 + 2t_5 + 3t_6,$$

$$N_+ = t_7 + t_8 + t_9 + 2t_{10} + 2t_{11} + 3t_{12}.$$

We have a system of four equations in twelve unknowns which must be natural numbers and must meet the following constraints:

$$t_1 \le N_-; \qquad t_7 \le N_+; \qquad t_2 \le N_- - t_1; \qquad t_8 \le N_+ - t_7;$$

$$t_3 \le N_- - t_1 - t_2; \qquad t_9 \le N_+ - t_7 - t_8;$$

$$t_4 \le (N_- - t_1 - t_2 - t_3)/2; \qquad t_{12} \le (N_+ - t_7 - t_8 - t_9)/3;$$

$$t_1 \le v_1; \qquad t_7 \le u_1; \qquad t_2 + t_4 \le v_2; \qquad t_8 + t_{10} \le u_2;$$

$$t_3 + t_5 \le v_3; \qquad t_9 + t_{11} \le u_3.$$

There always exists at least one, and at most a finite number of solutions of the four equations with these constraints. Each particular solution corresponds to a single form of a minimal subgraph of the external SR-graph for atom A_i. All minimal external SR-graphs G_R^E for the $S(A) \Rightarrow S'(A)$ transformation are obtained as all possible combinations of all subgraphs of the external SR-graphs of each atom $A_i \in S(A)$. The implemented algorithm generates all external SR-graphs $G_R^E = (V_R^E, R_R^E, \emptyset, \psi^E, \{-1, 1\})$, and is as follows:
1. Set $V_R^E = E_R^E = \emptyset$, $m = |A|$.
2. Find all solutions $T^i = \{T^i_1, T^i_2, ..., T^i_m\}$ for all atoms $A_i \in S(A)$, where $T^i_j = (t_1, t_2, ..., t_{12})$.
3. Loop in i over all possible permutations of T^i.
4. Loop in j over all atoms in $S(A)$.
5. Let $t = (t_1, t_2, ..., t_{12}) = T^j_{u_{ij}}$.
6. Loop in k over 1 to 12.
7. Loop in s over 1 to t_k.
8. $m = m + 1$, $V_R^E = V_R^E \cup \{v_m\} \cup \{v_j\}$ (v_m for $m > |A|$ is a new virtual atom).
9. $e \{v_j, v_m\}, E_R^E = E_R^E \cup \{e\}$, if $k \le 6$ then $\psi^E = -1$
$$\text{else } \psi^E = 1$$
10. If $(k > 3) \wedge (k < 7) \wedge (k > 9)$ then $E_R^E = E_R^E \cup \{e\}$.

11. If $(k = 6) \wedge (k = 12)$ then $E_R^E = E_R^E \cup \{e\}$.
12. End of s cycle.
13. End of k cycle.
14. End of j cycle.
15. $\mathfrak{R} = \mathfrak{R} \cup G_R^E$, $V_R^E = E_R^E = \emptyset$.
16. End of i cycle.

We say, that the SR-graph G_R fulfills the star condition [20] (denoted (∗)) if

1. it contains no loop.
2. it is an ∈-graph, i.e. it is a graph that contains an Euler path when isolated vertices are omitted,
3. for each vertex, the absolute value of the sum of evaluation of all edges incident with this vertex is less than 2, and
4. the sum of evaluation of all edges is equal to 0.

4.2.2 Reaction Distance

The algorithm for the computation of the reaction distance is one of the most important algorithm of the synthon model of organic chemistry. The *Principle of Minimal Reaction Distance* is a reaction distance analog of the Ugi's *Principle of Minimal Chemical Distance* and is used as the main heuristic to reduce the number of transformations employed, and synthons generated. The efficiency of the algorithm for the computation of the reaction distance is crucial for unable implementation of the whole synthon model.

The reaction distance between two isomeric synthons $S(A)$ and $S'(A)$ is computed as follows:

1. Construct the SR-graph G_R corresonding to the transformation $S(A) \to S'(A)$.
2. Let $D = 0$.
3. Decompose $G_R = G_1 \cup G_2 \cup G_3$ so that
   - G_1 contains all vertices with one loop at least, and all incident edges.
   - G_2 contains all virtual vertices, and all incident edges, unless they are in G_1.
   - G_3 contains all remaining vertices and edges.
4. Decompose $G_1 = G_1^+ \cup G_1^-$ so that G_1^+ contains only edges from G_1 with the mapping $+1$ and loops with the mapping -1, and G_2^- contains edges and loops with opposite mappings.
5. For each graph
   $G_i^+ = \{\{v_1, v_2\}$, one electron l_1 on v_1, one electron l_2 on v_2, $\{e_1 = \{v_1, v_2\}\}$, $\{e_1 \mapsto +1, l_1 \mapsto -1, l_2 \mapsto -1\}\} \subset G_1^+$
   which is not a subgraph of $G_i' = \{\{v_1, v_2, v_3\}, \{$ one electron l_1 on v_1, one electron l_2 on v_2, one electron l_3 on $v_3\}, \{e_1 = \{v_1, v_2\}, e_2 = \{v_2, v_3\}\}, \{e_1 \mapsto +1, e_2 \mapsto +1, l_1 \mapsto -1, l_2 \mapsto -1, l_3 \mapsto -1\}\} \subset G_1^+$, let $D = D + 1$, and remove G_i^+ from G_1^+.
6. For each analogical graph $G_i^- \subset G_1^-$ remove G_i^- from G_1^- and let $D = D + 1$.
7. For each vertex $v \in V_1^+$ which is incident to exactly one edge and at least to one two-electron loop, remove this edge and one loop, and let $D = D + 1$.
8. Repeat this process with vertices of the graph G_1^-.
9. Create a new graph $G' = G_1^+ \cup G_1^- \cup G_3$ and decompose it to $G_1' \cup G_3'$, where G_1' and G_3' have the same properties as G_1 and G_3 above, respectively.

10. Create a new graph $G'' = G'_1 \cup G_2$.
11. Loop over all connected components $G^J \subset G'_3$.
12. Let $N_1 = |V^J \cap V'_1|$, $N_2 = |V^J \cap V_2|$.
13. If $(N_1 > 1) \vee (N_1 + N_2 > 2 \vee N_2 \cdot (N_1 + N_2) = 2$ then try the next component.
14. If G^J fulfills the star condition (∗), remove it from G'_3, let $D = D + |E^J|$, and try the next component.
15. If $N_1 + N_2 = 0$ and there exists $h \in E^J$ such that $G^J - \{h\}$ fulfills the star condition (∗), remove G^J without the edge h from G'_3, let $D = D + |E^J|$, and try the next component.
16. If $N_1 + N_2 \neq 0 \wedge N_1 = 1$, then find the only common vertex k of graphs G'_1 and G^J. If k is incident to only one edge h and if $G^J - \{h\}$ fulfills the star condition (∗), then remove G^J from G'_3, let $D = D + |E^J|$, and try the next component.
17. If $N_1 + N_2 \neq 0 \wedge N_1 \neq 1$, and if in G_2 exists an edge h incident to some vertex of G^J such that $G^J \cup \{h\}$ fulfills the star condition (∗), the remove G^J from G'_3, h from G_2, let $D = D + |E^J| + 1$, and try the next component.
18. End of the loop.
19. From G' remove each edge $h = \{v_1, v_2\}$ such that
    - $v_2 \in V_2$,
    - v_1 is not incident to any loop,
    - there does not exist any edge in $G' \cup G'_3$ incident to v_1, which has the sign opposite to the sign of the edge h.
    Let $D = D +$ number of all removed edges.
20. Let $G' = G' \cup G'_3$, $D = D + |E'|$.
    The non-combinatorial part of the algorithm ends here. In the following, each edge in G' is substituted by one combination of
    - two one-electron loops, one on each vertex of the edge,
    - one two-electron loop on the first vertex of the edge, or
    - one two-electron loop on the second vertex of the edge.
    (Multiple edges are taken as several single edges, and the sign of the edge is retained in loops.)
    There will be $3^{n_1} 5^{n_2} 7^{n_3}$ such substitutions, where n_1, n_2, and n_3 are numbers of single, double, and triple bonds, respectively. For each substitution a number $N = A^+ + A^- + B^+ + B^- - C_1 - C_2 - E_1 - E_2$ is computed, where
    A^+ is the number of two-electron loops which map to $+1$ in $G' - G_2$;
    A^- is the number of two-electron loops which map to -1 in $G' - G_2$;
    B^+ is the number of one-electron loops which map to $+1$ in $G' - G_2$;
    A^- is the number of one-electron loops which map to -1 in $G' - G_2$;
    $C_1 = \min(A^+, A^-)$;
    $C_2 = \min(B^+, B^-)$;
    $E_1 = \min(F(A^+ - A^-), F(B^- - B^+) \div 2)$;
    $E_2 = \min(F(A^- - A^+), F(B^+ - B^-) \div 2)$,
    and $F(x)$ is defined above.
    Then let $D = D + N_{\min}$, where $N_{\min}$ is the minimal N computed.

169

As may be easily seen that the program package PEGAS uses the simplified version of the reaction distance algorithm, where no remapping of synthons is searched for before the computation of the reaction centre between $S(A)$ and $S'(A)$ (cf. section 2.3).

4.2.3 Reaction Centre Transformation

The first step in SPS generation is the generation of all synthetic precursors/successors of the specified reaction centre. As the number of atoms in the reaction center is usually small (often no more then 3 atoms), a simple combinatorial structure generation algorithm [3, 52] may be used.

To suppress the possibility of combinatorial expansion, algorithm GENFIS uses the table of valence states as the source of stable valence states for each atom of the reaction center. Let the set of stable valence states of atom $x_i \in X$ for the reaction centre $S(X)$ be $\mathscr{D}_i$. The algorithm GENFIS for the generation of all synthetic precursors/successors of the reaction center consists of the recurvise algorithm GENFIS1 which generates all transformations of $S(X)$ such that each atom $x_i \in X$ is in the specified valence state $v_i \in V$, and a simple driver which is responsible for the generation of all permutation of stable valence states for all atoms in X.

The GENFIS1 algorithm is as follows:
1. If $|X| = 1$ then output the synthon $S(x)$ where x ($\{x\} = X$) is in valence state $v \in V$.
2. If $|X| > 2$ then go to step 8.
3. Generate all possible combinations of atoms x_1 and x_2 in valence states $v_1 = (0, w_1, w_2, w_3)$ and $v_2 = (0, u_1, u_2, u_3)$, respectively. This is performed by the following steps:
4. If $w_1 > 0$ and $u_1 > 0$ then connect both atoms with the single bond, and add the synthon to the output set.
5. If $w_2 > 0$ and $u_2 > 0$ then connect both atoms with the double bond, and add the synthon to the output set.
6. If $w_3 > 0$ and $u_3 > 0$ then connect both atoms with the triple bond, and add the synthon to the output set.
7. Add both atoms disconnected to the output set, and the algorithm.
8. Remove arbitrary atom α from the set X, and call recursively the algorithm GENFIS1 on the resulting synthon Y. Let the output of this call be denoted $\mathscr{L}$.
9. For each synthon $S(Y) \subset \mathscr{L}$ generate all possible combinations with the atom α in analogy to steps 4–7 above.
10. The set of all obtained synthons (without isomorphic duplicates) is the output of the algorithm.

The whole GENFIS algorithm not only permutes all valence states of all atoms in the reaction center, it also checks for the fulfillment of the star condition (∗):
1. Let $\mathscr{R} = \emptyset$ be the set of SPS of the reaction center $S(X)$.
2. For each permutation V_j of valence states $v_{ij} \in \mathscr{D}_i$ for each atom $x_i \in X$ loop.
3. Call GENFIS1 $(S(X), V, \mathscr{N})$.

4. Check each synthon from $\mathcal{N}$ against the (*) condition, and add to $\mathcal{R}$ only those which passed the test.
5. end loop.

4.2.4 Stabilization and SPS Generation

In principle, it is possible to use the GENFIS algorithm for the generation of the SPS of any synthon. In practice, as was explained in the preceeding sections, this must lead to a combinatorial explosion for all but trivial tasks. The stabilization permits us to generate directly the synthetic precursors/successors of any given synthon. The $STAB(S(X), S(A), \mathcal{N})$ is used to find all stabilizations $S'(A) \in \mathcal{N}$ of synthon $S(X)$ with respect to $S(A)$ ($X \subset A$):
1. If $|X| = |A|$ then $S'(A) = S(A)$ is output.
2. Find the set $Y \subset A$ of all atoms from the first neighborhood of X.
3. Find the set $C \subset Y$ of atoms whose valence state must be changed in order to connect them to (or disconnect from) atoms in $S'(X)$. There may be at most 2 such atoms.
4. If $|C| = 0$ (there is no such atom), immerse the $S'(X)$ to $S(A)$ and the resulting synthon $S'(A)$ is output.
5. For each $c_i \in C$ perform all permitted changes of its valence state and connect all atoms from Y to $S'(X)$ obtaining $S'(X \cup Y)$ (all atoms from Y with expect of atoms from C may be connected to $S'(X)$ without change of their valence state).
6. For each synthon $S'(X \cup Y)$ call $STAB(S'(X \cup Y), S(A), \mathcal{N}_1)$ and put $\mathcal{N} = \mathcal{N} \cup \mathcal{N}_1$.

Combination of all the above presented algorithms (and many others, especially the algorithm for the generation of the canonical code of a synthon, the algorithm for the computation of chemical distance between two synthons, etc.) gives the final algorithm for SPS generation:
1. Input the synthon $S(A)$, its reaction centre $S(X)$, and its passive part $S(\bar{X})$.
2. Generate all transformed reaction centres $\mathcal{R} = \cup S'(X)$.
3. Let $\mathcal{M} = \emptyset$ be the set of SPS of $S(A)$.
4. Call $STAB(S(A \setminus \bar{X}), S'(X), \mathcal{N})$ for each synthon $S'(X) \in \mathcal{R}$, and add each synthon $S'(A \setminus \bar{X}) \in \mathcal{N}$ which could be immersed to the synthon $S(A)$ to the set $\mathcal{M}$.

5 Illustrative Examples of Tasks Performed by PEGAS

The program PEGAS was worked out for CAOS (Computer Aided Organic Synthesis), it means that it serves as a supporting tool for the intentioned fantasy of synthesizing organic chemists. It produces retrosynthetic trees, facilitating choice of effective ways toward a target compound.

Plausible disconnections of structures and recombination of dissociated parts (synthons) must take into account the quality of potential reaction centers, which could serve for dissociation or association. By quality of reaction centers, it can be understood as the preferable type of charge on the atom, given by the type of dissociation of a bond by heterolysis or homolysis, from which follows the affinity of reactants in spe (i.e. Nu/E, R/R etc.). The PEGAS system can operate with all realistic valence states of elements incorporated into a small database (i.e. with charged atoms, radicals, valence states typical for intermediates and transition states etc.). Within reaction paths product from one stable compound to another, parts of different reaction mechanisms can be found. The quality of reaction centers, i.e. evaluation of possibilies, extent and consequences of interactions of chosen reaction centers should be included in the subsequent version of PEGAS, which is in preparation (i.e. PROPOS).

However, it is the user's decision as to which valence states of which elements should be included in the ensemble of FVS: metals can also be involved. So, organometallic, and to some extent, inorganic chemistry can be studied.

Planning steps and reaction mechanisms of forward reactions (from educts to products) as well as of retroreactions (from target compound through intermediates to educts) can be performed by PEGAS

The examples contain:
— generation of possible isomers,
— proposals of reaction mechanisms:
   possible mechanisms of reaction of cyclopropanamine with hypochlorite
   close view of the Peterson olefination reaction
   generating of proposals of mechanisms of reactions with reactants in less common valence states
   mechanisms of radical reactions
   proposals of molecular rearrangements mechanisms
   generation of systematic mechanistic hypotheses for reactions represented by Reaction Cubes, and
— planning of retrosynthesis:
   methods of forming a benzthiazepinone cycle designed by PEGAS

5.1 Generation of Possible Isomers

The diazomethane molecule (A) in spite of containing only five atoms, has six structural isomers (B–G) and five resonance structures (1–5). Of its structural isomers, diazirine (B) and cyanamide (C) are stable at room temperature, and isocyanamide (D) and carbodiimide (E) are known in the form of stable derivaties; isodiazirine (F) has not yet been described in the literature and transition derivatives of nitrilimine (G) were reported in 1959 [53].

Nitrilimines can be obtained as transient systems of, e.g., dehydrohalogenation of hydrazonoylhalides. Betrand et al. [54] produced them from lithium salts of diazo compounds (Eq. 42).

$$R-\overline{C}^- = N^+ = \overline{N}^- \underset{Li^+}{\longrightarrow} R-C\equiv N^+ -\overline{N}^- -E \longrightarrow \overset{R}{\underset{E}{\diagdown}} = \overline{N}-\underline{\overline{N}} \qquad (42)$$

Program PEGAS generated these structures as potential precursors of nitrilimines. Other possible intermediates are mentioned in an abbreviated survey (Scheme 8).

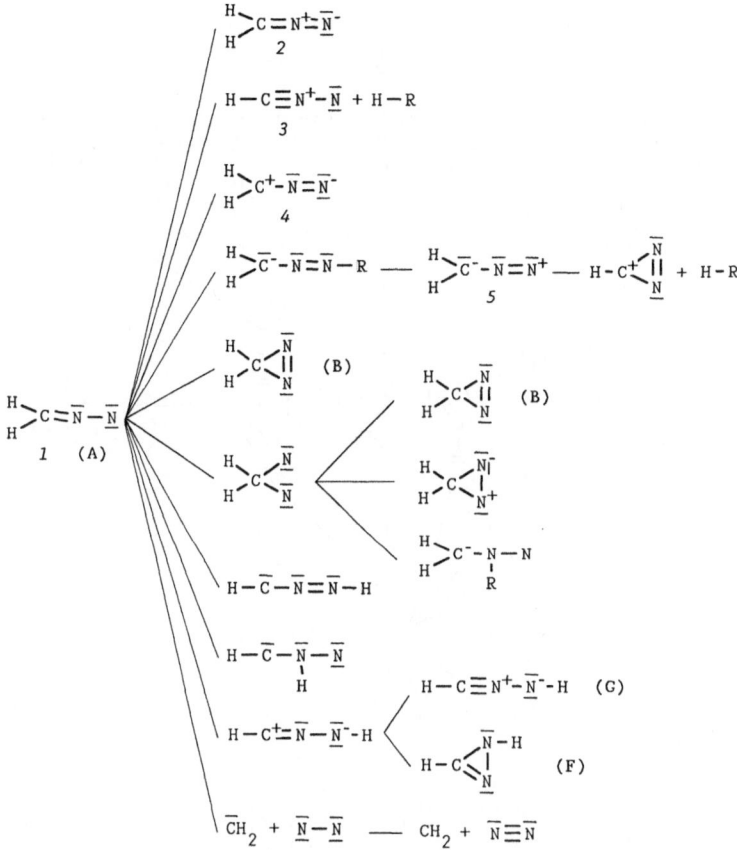

Scheme 8.

Isomers C, D, and E were generated by the program only on the fourth and fifth level from structures F and G. Because the program does not contain any heuristic forbidden systems with a triple bond in the trigonal cycle, this system (in square brackets in the scheme) was also actualized for further generation and served as a precursor of six other potential isomers including structural isomer D (Scheme 9).

5.2 Proposals of Reaction Mechanisms

5.2.1 Possible Mechanisms of the Reaction of Cyclopropanamine with Hypochlorite

Amino acids react with HOCl by decarboxylation producing imines [55]. The sequence of valence changes for 1-aminocyclopropancarboxylic acid was produced by program in the following way:

173

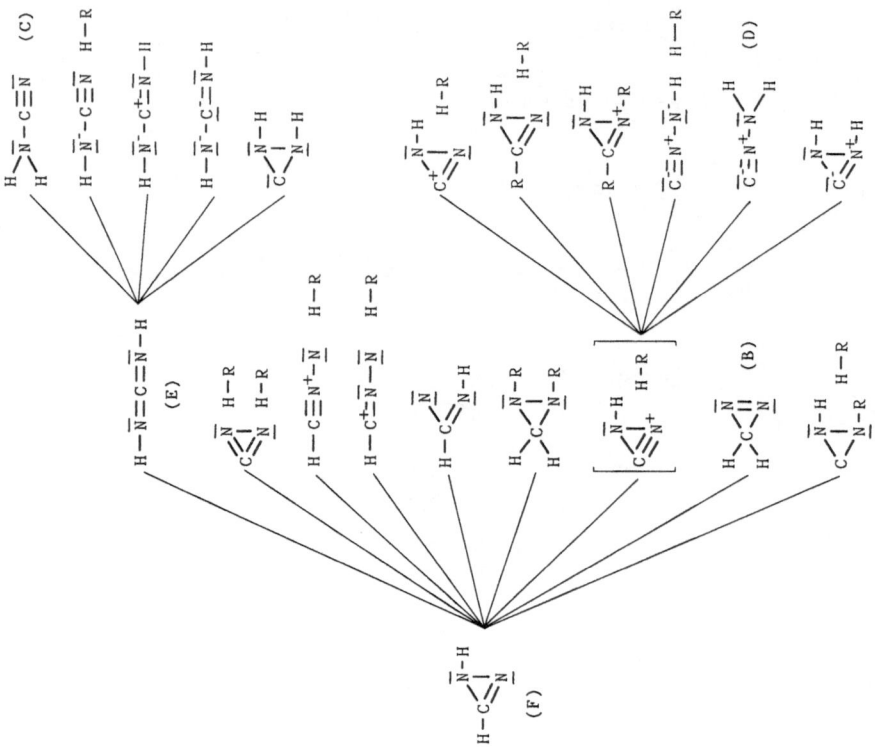

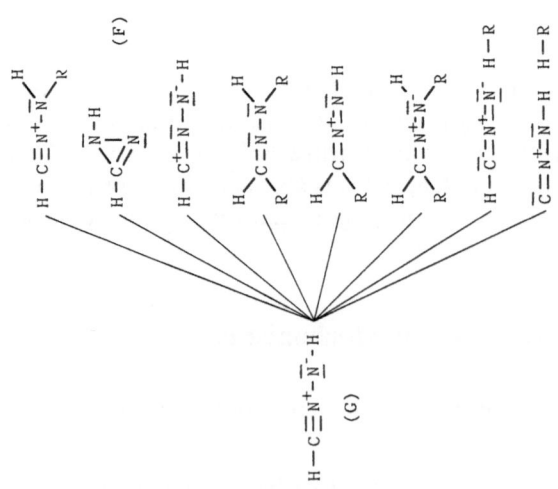

Scheme 9.

The key structure *1a* was generated by electrophilic substitution (Eq. 43), and then used as a starting structure for PEGAS.

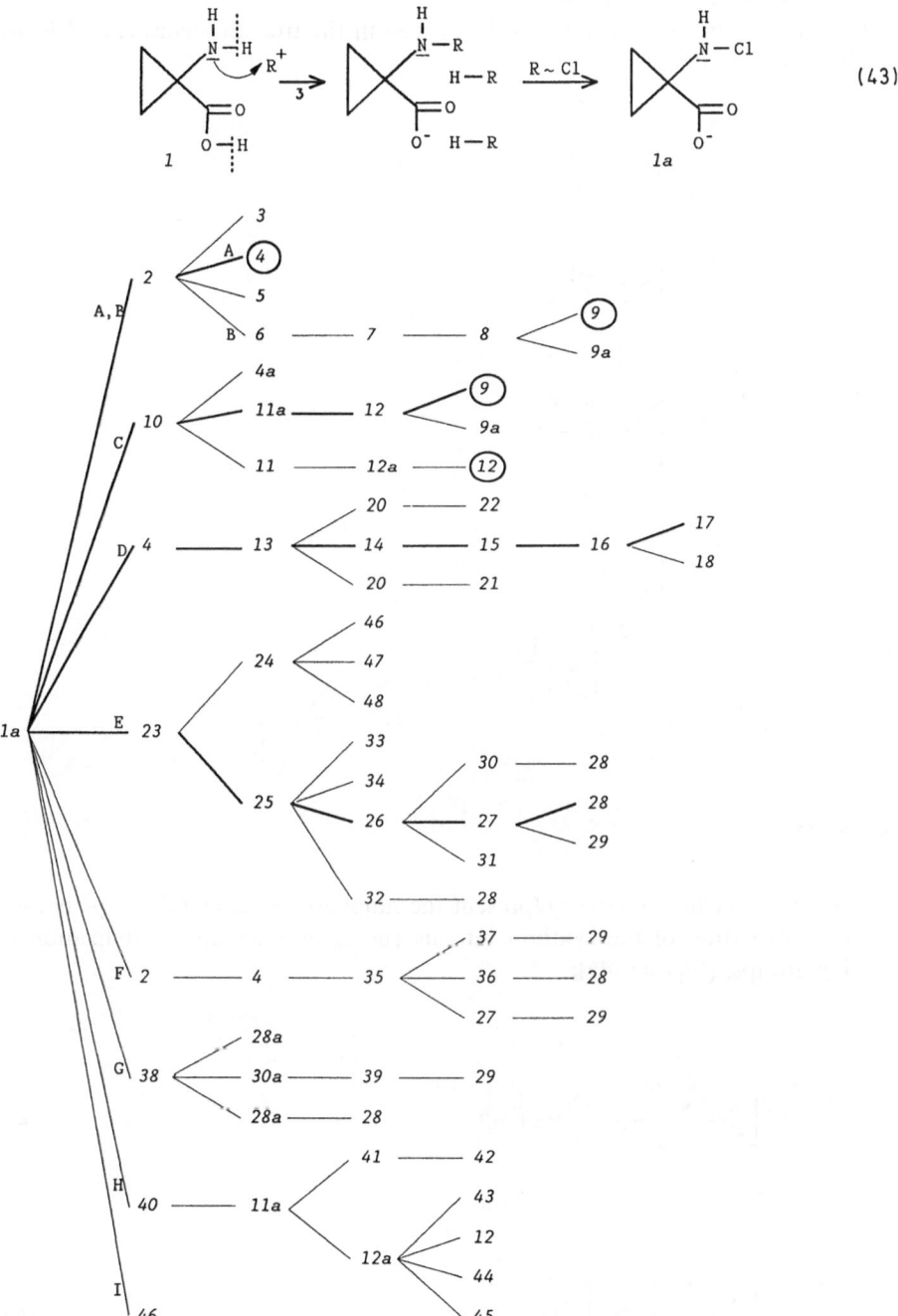

Scheme 10.

As a part of generated reaction tress (as an actual reaction center atom C1 was chosen, and program options IONS, NON-REDOX were used, cf. Sect. 4.1.4), 48 synthons were selected (Scheme 10).

Branches paths denoted by A to I are seen in the tree. Branches A and B are analyzed in Scheme 11.

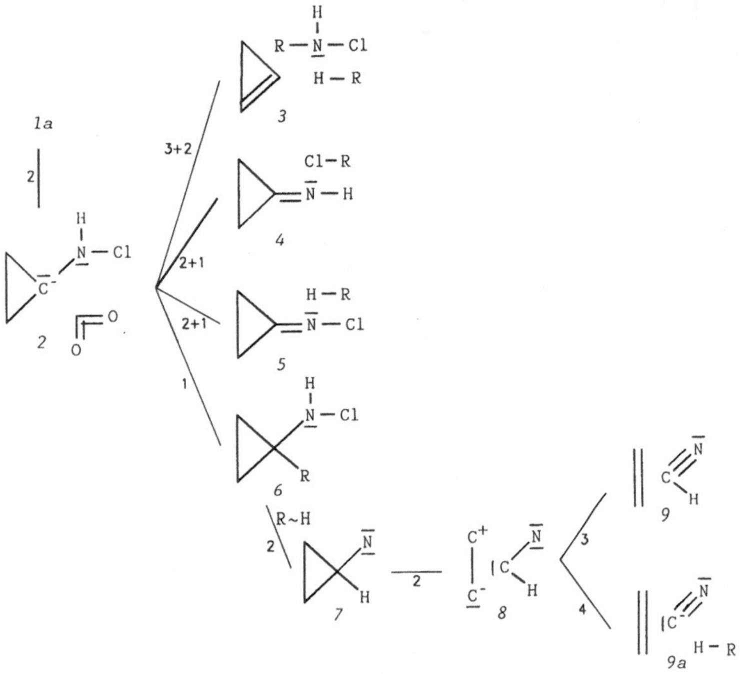

Scheme 11.

Numbers attached to edges represent the numbers of elementary steps needed for transformation of the synthon into its successor (including stabilization of leaving groups, (Eqs. 44–46):

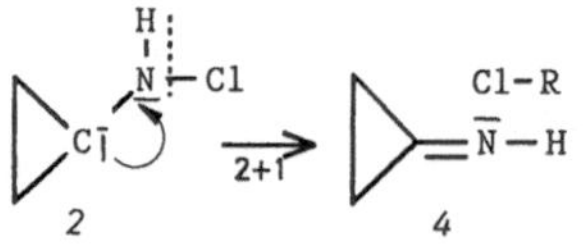

$$(44)$$

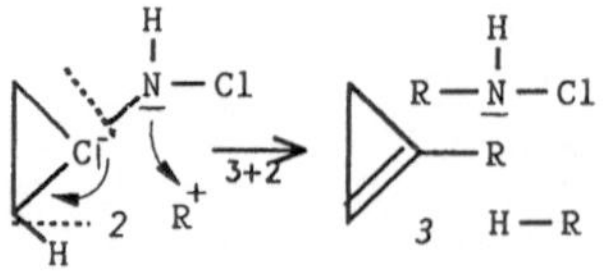

$$(45)$$

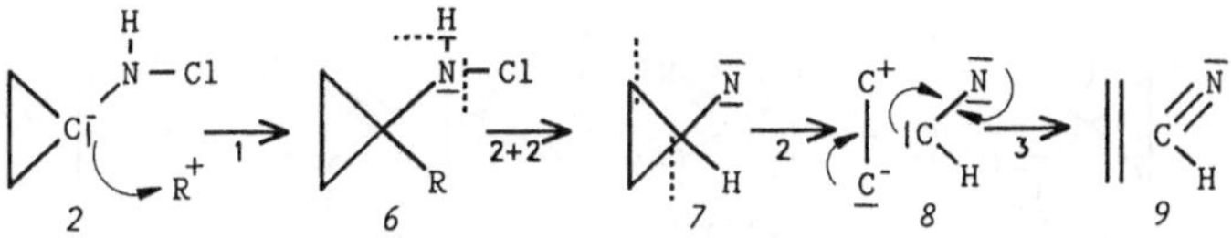

$$ (46) $$

The sequence $1-1a-2-4$ is labeled as a branch A: Synthon 4 is considered as stable, although in real chemical reactions this compound appears as an intermediate.

The sequence $1a-2-6-7-8-9$ (or $9a$) is labeled as a branch B: the successive sequence of changes of educt edges terminates at the stable synthons 9 ($9a$); as can be seen from synthon 8, the program works with valence states of charged atoms and offers quasi-structures in their extreme forms.

On the tree of reactions in Scheme 12, another branch (C) can be studied:

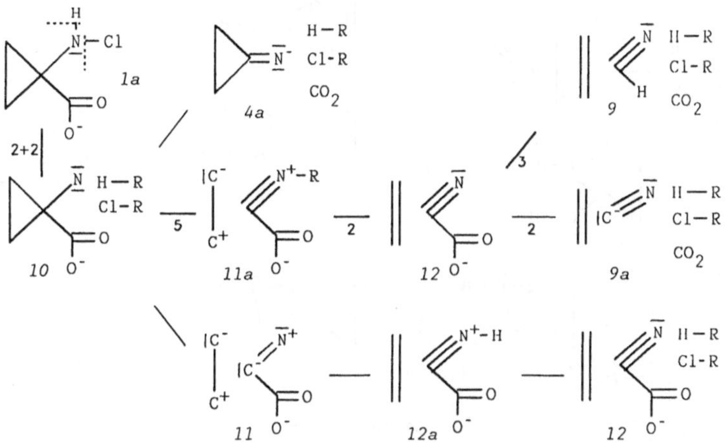

Scheme 12.

Elementary steps of reorganization of valence electrons have the following sequences (Eq. 47):

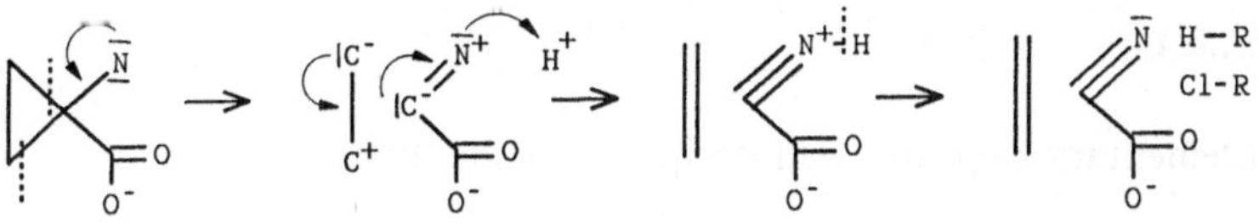

$$ (47) $$

Within steps $10-11a-12-9$, the following internal steps are latently present (Eq. 48):

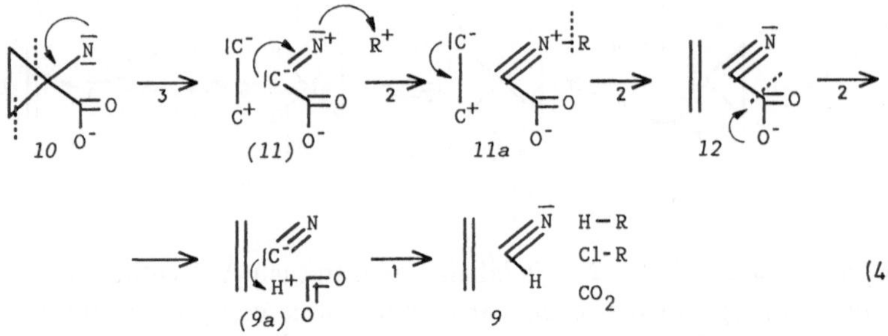

$$
10 \longrightarrow_{3} (11) \longrightarrow_{2} 11a \longrightarrow_{2} 12 \longrightarrow_{2}
$$

$$
\longrightarrow (9a) \longrightarrow_{1} 9 \quad
\begin{array}{l} H-R \\ Cl-R \\ CO_2 \end{array} \tag{48}
$$

The branch D corresponds to a generation of synthons produced by interaction of imine *4* with reagent $Cl-O-$ giving synthon *13* (Scheme 13):

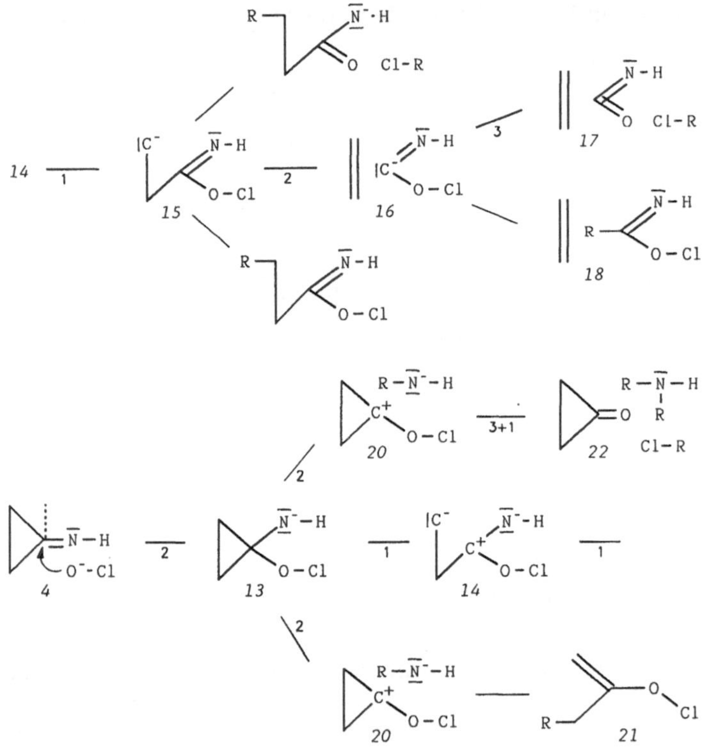

$$
14 \longrightarrow_{1} 15 \longrightarrow_{2} 16 \longrightarrow_{3} 17
$$

$$
13 \longrightarrow_{2} 14 \longrightarrow_{1} \quad 20 \xrightarrow{3+1} 22
$$

$$
4 \longrightarrow_{2} 13 \longrightarrow_{2} 20 \longrightarrow 21
$$

Scheme 13.

Elementary steps are well deciphered (Eqs. 49 and 50):

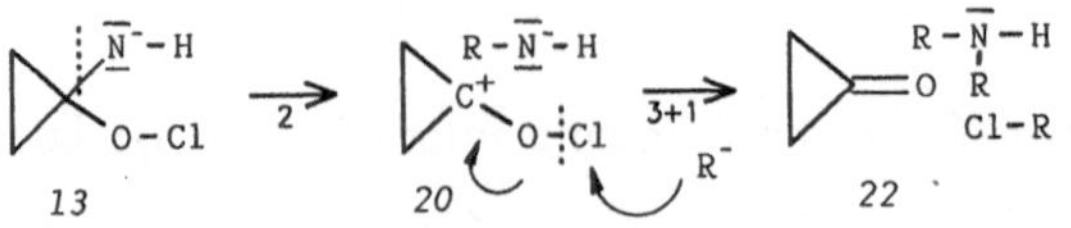

$$
13 \longrightarrow_{2} 20 \xrightarrow{3+1} 22 \tag{49}
$$

$$(50)$$

A substitution product can be generated from synthons *1* and *15* as well as from ethylene and isocyanic acid (Eq. 51):

$$(51)$$

Path *13–20–21* contains less probable splitting of $C-C$ bond of cyclopropane system (Eq. 52).

$$(52)$$

The branch E of reaction tree of educt *1a* (Eq. 53) was studied up to the fifth level: It starts by decarboxylation (with exception of synthon *24*, whose ketene skeleton was generated only at the first level and represents a blind branch) (Scheme 14).

$$(53)$$

Branch F also represents a path towards synthons of the nitrile class: Branch E produces them via decarbonylation, branch F via decarboxylation (Scheme 15).

The sequence of steps presented assumes an electrofugal (from the point of view of the substituted atom C1) opening of the cycle for elimination and it forces the hydride anion to leave the nitrogen atom. This concerns not only acrylonitrile formation *29*, but also formation of beta-hydroxypropionitrile *28*.

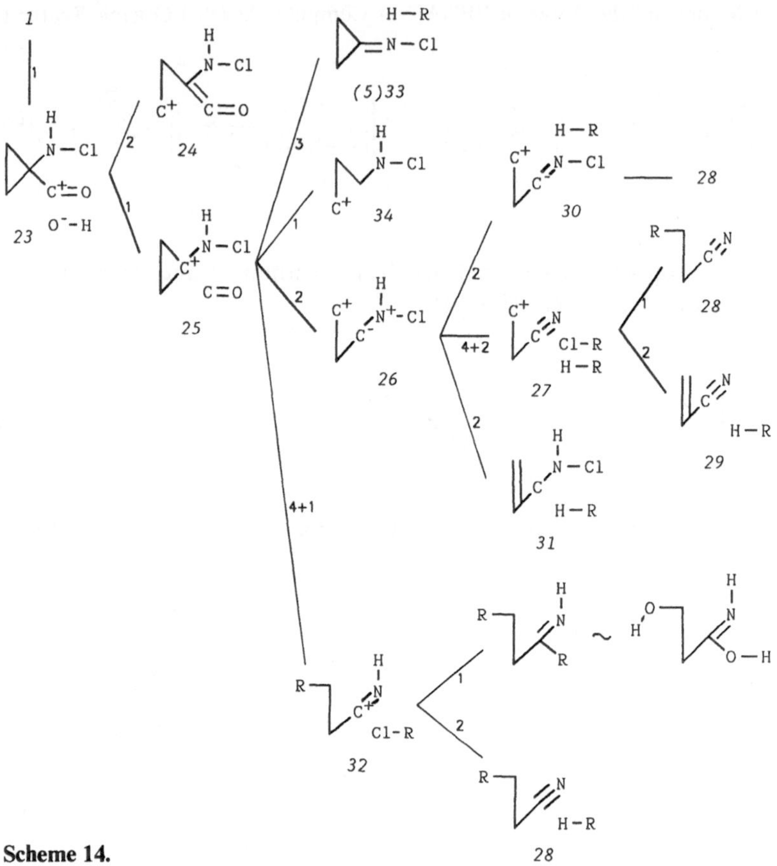

Scheme 14.

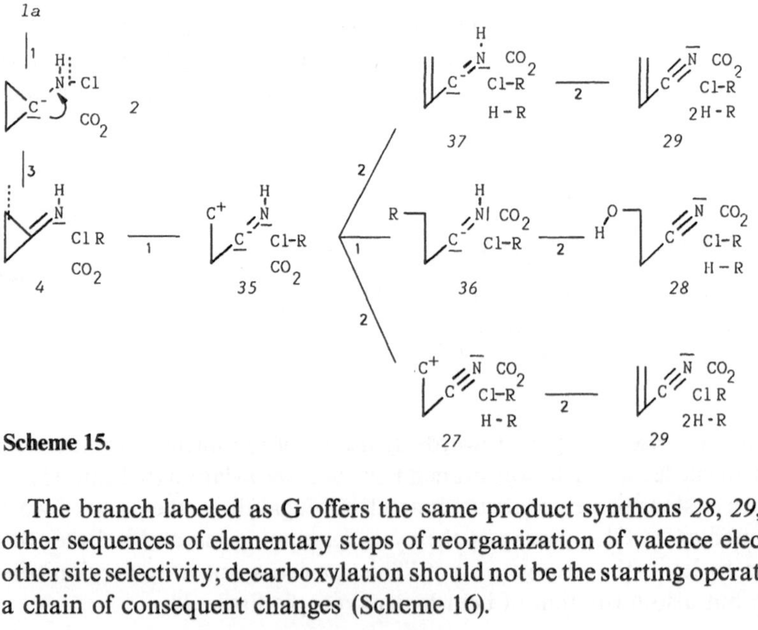

Scheme 15.

The branch labeled as G offers the same product synthons *28, 29*, obtained by other sequences of elementary steps of reorganization of valence electrons and by other site selectivity; decarboxylation should not be the starting operation, inducing a chain of consequent changes (Scheme 16).

180

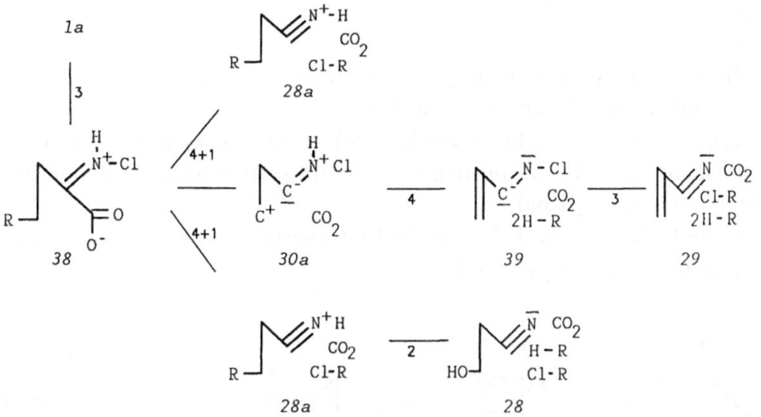

Scheme 16.

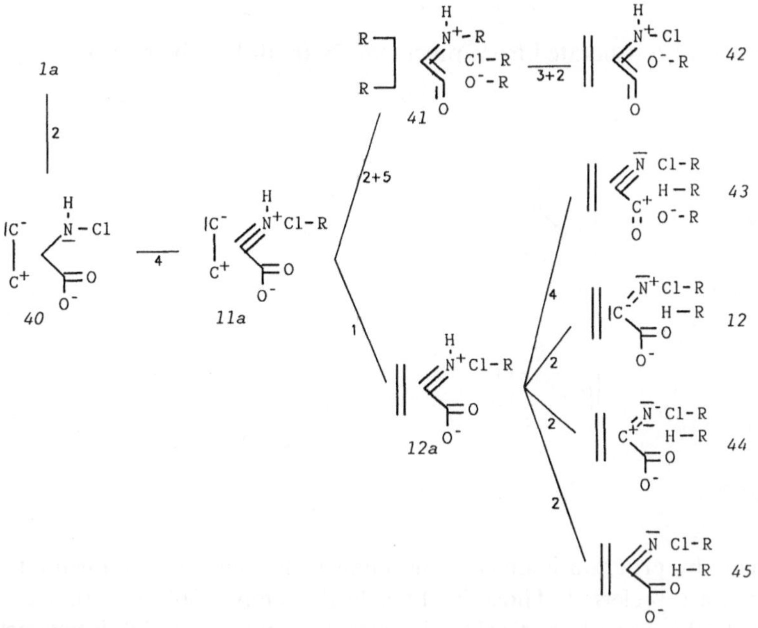

Scheme 17.

Sequence of elementary steps is like this (Eq. 54):

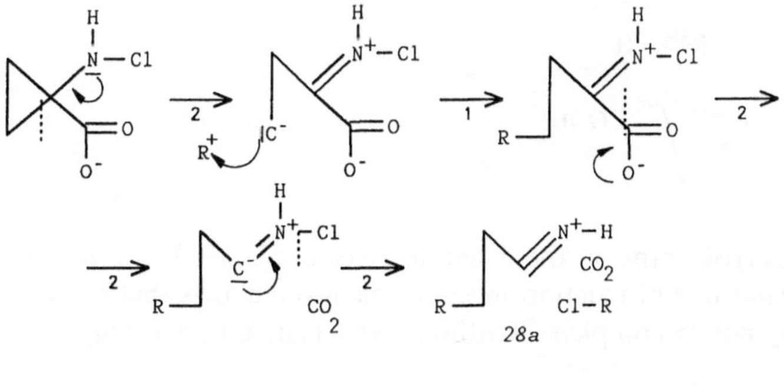

$$\tag{54}$$

Branch H shows fragmentation of synthon *la* onto ethylene, sometimes producing rather curious fragments (Scheme 17).

The path leading to mononitrile of oxalic acid *12* or to its anion need to necessarily lead via nitrene *10*, i.e. branch C, but the proposed splitting of the cycle via carbene *40* seems unusual.

The last of the branches given, I, has as its first synthon *46*, which is formed by the following sequence of steps (Eq. 55).

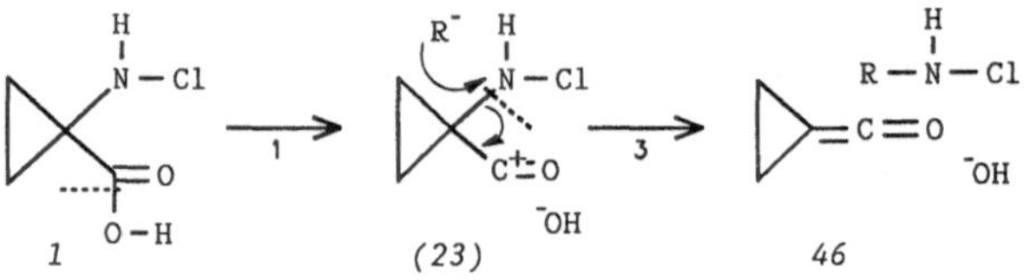

(55)

Synthon *46* was also generated from precursor *24* (path E, Scheme 18)

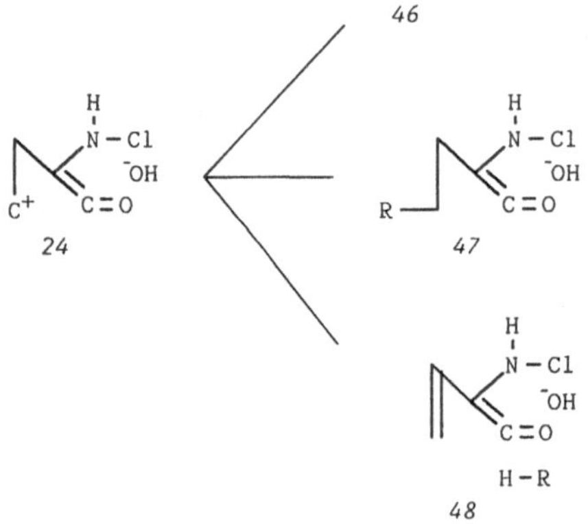

Scheme 18.

The offer made by program is diverse: mechanisms leading to experimentally proved synthons are preferred. Those in the solved example [56] are: ethylene, 1,2-disubstituted ethane, carbon dioxide, 3-hydroxypropanenitrile *28*, isocyanic acid *17*, hydrogen cyanide *9*, acrylonitrile *29*, and cyanoformic acid *12*.

As potential intermediates are in the paper [56] considered cyclopropanimine *4*, nitrene *10*, amide *13*, and nitrenium ion

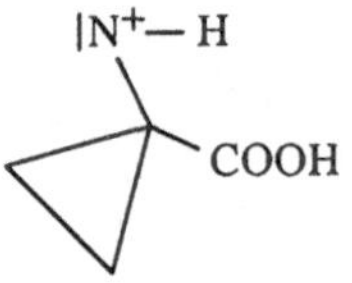

The preferred paths on the generated tree are labeled with heavy lines.

The partial tree of reaction mechanisms given shows that the proposals made in [56] may not be complete: Synthon *9* was generated not only via nitrene *10* by branch C, but also by branch B. 3-Hydroxypropanenitrile *28* was generated in three variants on the branch E, but also on branches F and G; similarly for acrylonitrile *29* four different possible ways of forming on branches E, F, F and G were proposed. Synthons *28* and *29* characterize group of compounds related to amides (they are not labelled in the schemes). The fact, that in mixtures not all components were identified allows us to suppose, that some other synthons generated by programs may not be improbable results (at least as transition states): e.g. cyclopropanone *22*, cyclopropene *3*, cyclopropylketene *39*, vinylketene *48*, and others.

The example given was evidence of the capacity of program PEGAS for proposing reaction mechanisms. (Radical and redox reactions were suppressed for simplicity).

5.2.2 Close-up view of the Peterson Olefination Reaction

According to [57] benzaldehyde reacts with bis(trimethylsilyl)methyl lithium giving mixture of *trans*- and *cis*-vinylsilanes PhCH = CHSiMe$_3$. The PEGAS system, from given educts with specified reaction centers, generated thirteen synthons (from the scheme, identical synthons and stabilizations of carbanion *1* by electrophiles are excluded). Their successive changes are also given in the reaction network (Scheme 19).

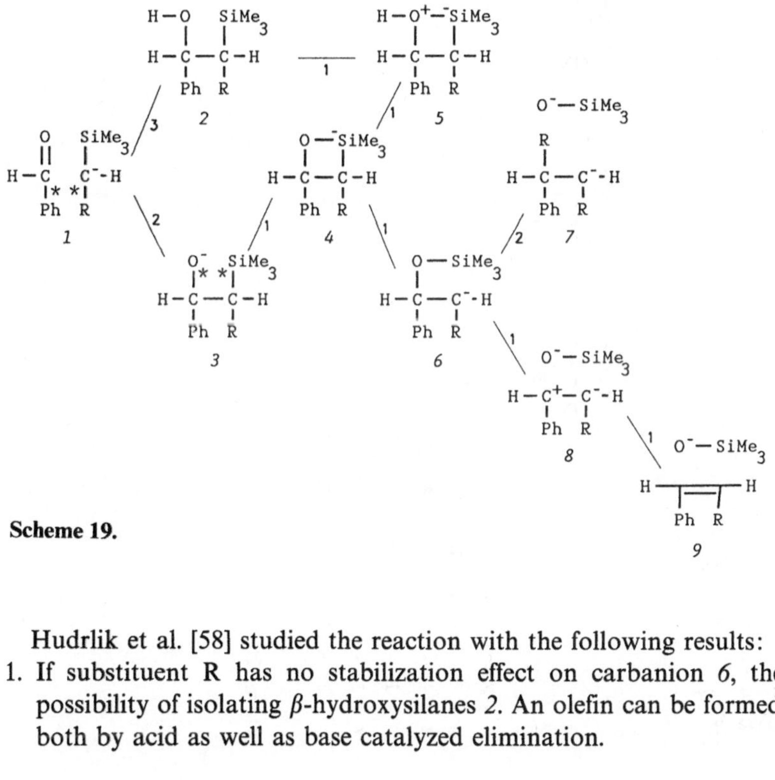

Scheme 19.

Hudrlik et al. [58] studied the reaction with the following results:

1. If substituent R has no stabilization effect on carbanion *6*, there exists a possibility of isolating β-hydroxysilanes *2*. An olefin can be formed from them both by acid as well as base catalyzed elimination.

2. When substituent R is electron-withdrawing, neither beta-hydroxysilane nor betaoxidosilane *3* can be isolated.
3. β-Elimination of β-hydroxysilanes under acidic conditions works by an *anti*-pathway and in basic conditions by a *syn*-pathway.
4. When point (2) is correct, the authors [58] proposed the formation of 4-centered species *4* as by Wittig reaction.

In the scheme produced by the PEGAS program there were proposed in detailed succession (up to one-step acts) all synthons, proposed by the authors of Ref. [58]: *2, 3, 4* (oxasiletan), *10*. Presumed simultaneous formation of C−C and Si−O

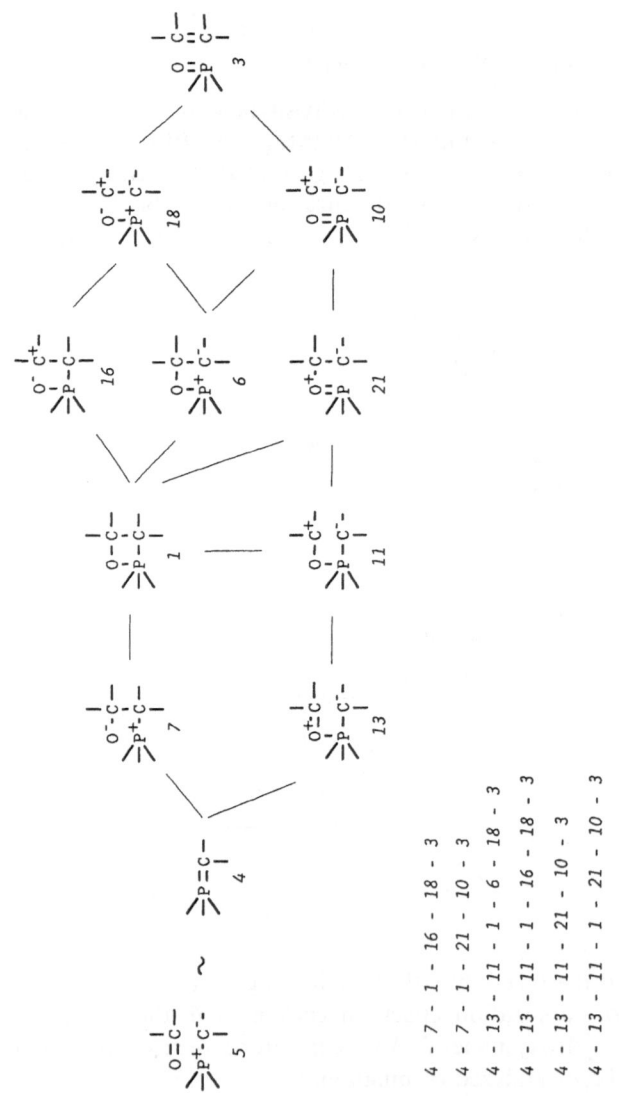

Scheme 20.

bonds is modeled in program by three-step process *1–3–4*, what is given by principle of elementary steps of reorganization of valence electrons [39].

When comparing the model of the mechanism of the Peterson reaction with the scheme of the Wittig reaction generated by PEGAS, we also can find in its output synthons which are analogous to synthons generated for the Peterson reaction (Scheme 20).

Structure *1* of oxaphosphetane is taken as proved. Ylid *5* or *4* as an intermediate stage on the way starting at educts $R_3P + Hal-CH_2-R'$ was generated as a precursor of the structure *1*; betaine *7*, which is produced by fast splitting of oxaphosphetanes by for instance lithium halides [59] can be considered as a successor of the systeme *1* in the opposite direction to the olefin route.

Synthons *13–11–1–16–18–3* and *13–11–21–10–3* model a mechanism which is assumed to be a start of interaction P ← O. Schneider's concept [60] presented as the first one a logical argument for stereochemical control in *cis*-selective Wittig reactions. An analogy of the structure *11* in the diradical can be found in the Olah study [61]. Despite the paper by Vedejs and Marth [62] some experimental evidence does not support the idea that the Wittig reaction is started by the interaction P ← O. The PEGAS program does not contain any built-in facts about reactions, and generated the presented possibility independently.

5.2.3 Generation of Proposals of Mechanisms of Reactions with Reactants in Less Common Valence States

The capacity of the program PEGAS is also satisfactory for processing of less common valence states of atoms. As an example, the proposal of the solution of deoxygenation of tetrahydrofuran by atomic carbon with an experimental and computational study by Shevlin et al. [63] was compared. Although the list of valence states does not differentiate between the basic state ^3P from the singlet excited states ^1D and ^1S, using the option FVS RADICALS, synthons were generated which correspond to the structures proposed by Shevlin.

The table of valence states of the carbon atom also contains VS 4000 (e.g. $\bar{C}$, $\cdot\bar{C}\cdot$), 3100 ($\cdot C-$) and 2010 ($C=$) among other more common valence states. The following proposals were generated (Scheme 21).

Results from the study [63] show that exclusive cleavage occurs in C_2H_4 and CO. The authors postulated a scheme of homolytic fragmentations of adduct *2* and assumed that the structures (using our numbering) were *2, 3, 13, 19*. Based on the computed energy, it can be concluded that these pathways may be opened in the deoxygenation of tetrahydrofuran by atomic carbon. The presence of either cyclobutane *21* or cyclopentanone *22* was not confirmed in experiments of these authors, although during the generation of structure *19* by other reactions, structure *21* is formed by cyclization. From structure *3*, 2,3-dihydropyran can be generated as an output of the formal insertion of C−O unit (in deoxygenation conditions [63] it was not proved).

In synthons with separate atom $\bar{C}$, i.e. *7–11*, in the following level of tree of reactions a unit $\bar{C}=$ is always created.

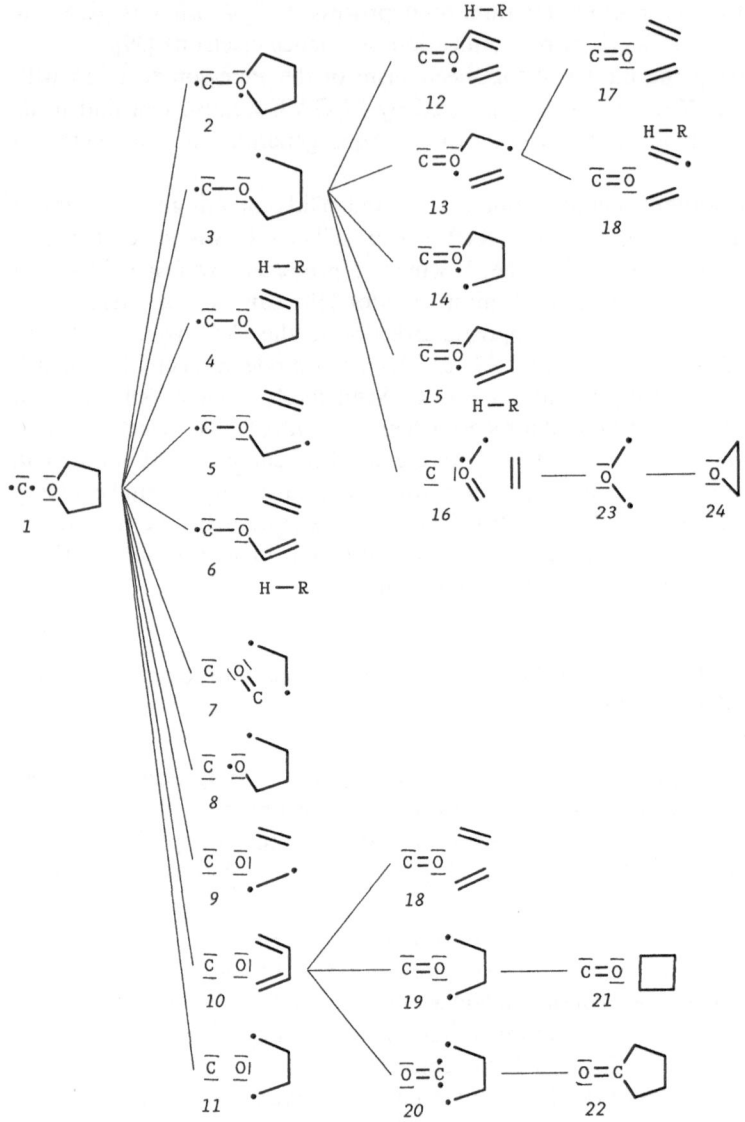

Scheme 21.

Since no conditions are given for starting the task, the program offers general solutions for a whole variety of reaction conditions and also gives the synthons 1,3-butadiene *10*, ethyleneoxide *24* etc.

5.2.4 Mechanisms of Radical Reactions

The PEGAS system is capable of formulating mechanisms of radical reactions including the participation of metals. The proposal for the mechanism of oxidative decarboxylation of 3-hydroxy carboxylic acids via V(V) complexes (Scheme 22) serves as an example.

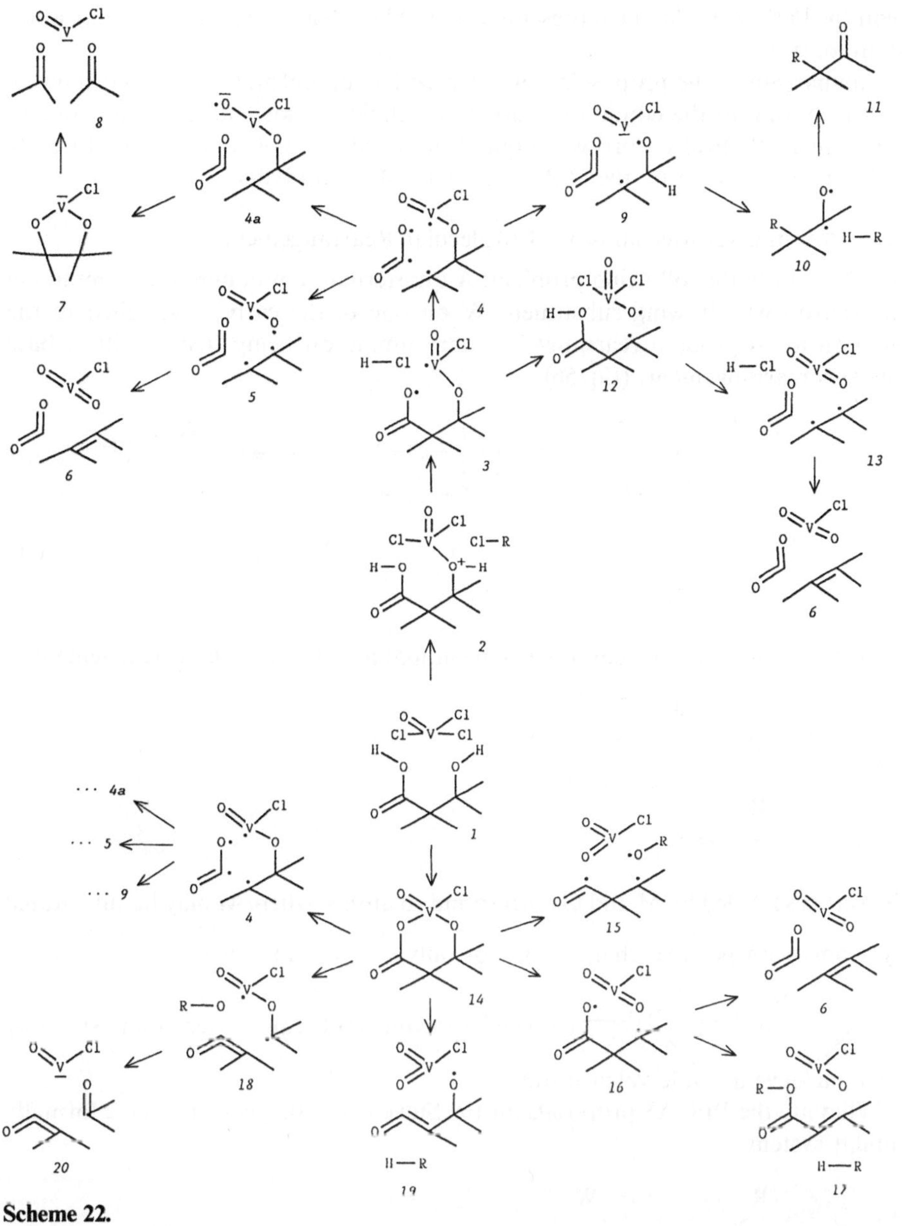

Scheme 22.

Olefin 6, aldehyde or ketone 8, acrylic acid 17 and ketene 18 can be found among the possible products generated; unfinished states 15 and 19 did not seem to be promising. Complex 14 and 1,4-metalla diradical 4 formed from it by decarboxylation is an important intermediate. The ways terminating at olefin contain path 4–5–6 starting with decarboxylation and path 2–12–13–6, where the decarboxylation follows after the homolysis of the C−O bond of the original hydroxyl. Complex 14 can homolyse into 4 or 16, in both cases diradicals produce olefin *6*. Path *4–9–10–11* represents a possible 1,2-shift leading to stabilization of diradical *9*.

Comparison of the proposals generated with experimental studies [64] showed, that according to the conditions, mostly by choice of solvent, it was possible to obtain in synthetically worthwhile quantities products *6*, *8*, and *11*. Structures *14* and *4* are considered by the authors of Ref. [64] as probable.

5.2.5 Proposals of Mechanisms of Molecular Rearrangements

We shall study the following problem: A quarternary ammonium salt containing an electron withdrawing substituent W on one of the carbons attached to the nitrogen atom yields a rearranged tertiary amine on being treates with a base (Stevens rearrangement) (Eq. 56).

$$(56)$$

In the graph of the reaction mechanism [65] assigned to this rearrangement

the vertices labeled by M and G correspond to atoms, where M may be substituted by atoms with positive charge (e.g. not only $\diagdown_{N^+}\diagup$ but also $\diagdown_{S^+}\!\!-\!,$ $\diagdown_{P^+}\diagup,$ $\diagdown_{As^+}\diagup,$ $\diagdown_{Se^+}\!\!-\!,$ $\diagdown_{Si^+}\!\!-\!,$ etc.) and applying a subsequent elementary step $1D_E$ it can achieve a stable valence state.

What are the PEGAS proposals for the Stevens rearrangement when a formally similar system

is treated not under the conditions of a heterolytic process but under conditions supporting homolysis/colligation (in a similar way to the diradical mechanism suggested by Bates [66] for the Stevens and Wittig rearrangements)?

Proposals offered by PEGAS are displayed in Scheme 23.

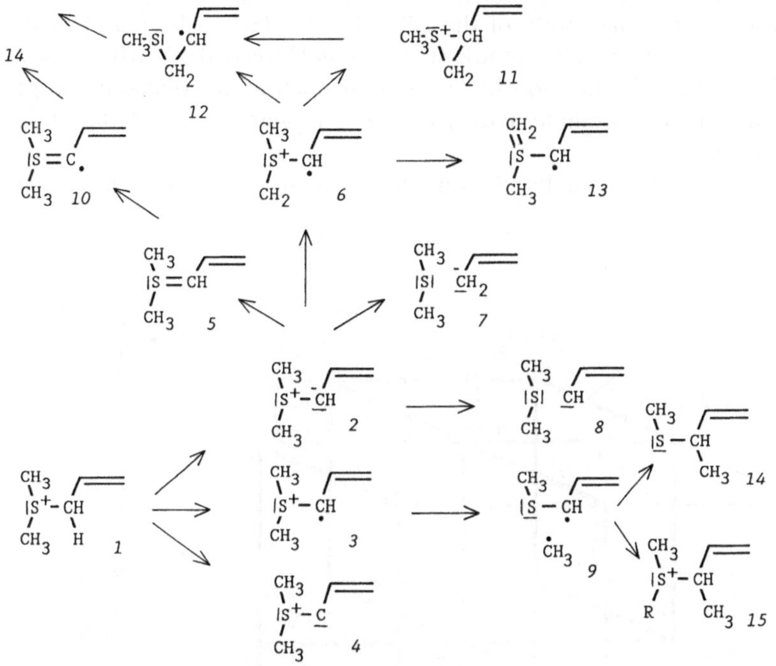

Scheme 23.

A proposal represented by the path *1–3–9–14* is an analog of the Stevens rearrangement of quarternary ammonium salts including also a similar solvent cage effect *5–10–14*. PEGAS did not omit an appearance of synthon *2* for the heterolytic process; in addition of carbenoid structures *4* and *8*, interesting proposals of rearrangements with homologisation *6, 11, 12* are also presented. A proposal of synthon *15* (after a rearrangement with subsequent alkylation of the sulfur atom) seems to be excessive.

The path *1–3–9–14* may be assigned to a mechanism of the photo-Stevens rearrangement of 9-dimethylsulfonium fluorenylide [67].

5.2.6 Generation of Systematic Mechanistic Hypotheses for Reactions Represented by Reaction Cubes

More O'Ferrall diagrams for the analysis of complex reactions (for a review see Scudder [68]). The progress cube of Grunwald as a quantification and extension of 2-dimensional More O'Ferrall (MOF) projections, allows a simple analysis of mechanisms involving three concerted reaction events. The ability of the PEGAS system to find automatically reactants, products and intermediate charged structures occupying reaction cubes edges was examined on a test example:

A reaction cube from PEGAS is generally constructed by placing neutral educts into one corner of the cube. Accessory reagents (e.g. HA, B⁻, ...) need not be

Eva Hladka et al.

introduced, program will generate them when necessary as virtual, closely unspecified atoms (HB as H−∗ etc.). Product(s) of the reaction decomposable into 3 processes (any of them made at the most of 3 ESRE) will be placed in the opposite corner of the solid body of the cube. Product(s) can be found in the production by the PEGAS system in the first and second level of the (retro)synthesis tree. The cube is placed in the coordinate system in such a way, that single edges are placed parallel to coordinate axes. Every axis represents one of the above mentioned processes.

For a further couple of analyzed examples only one common scheme was constructed (Fig. 15).

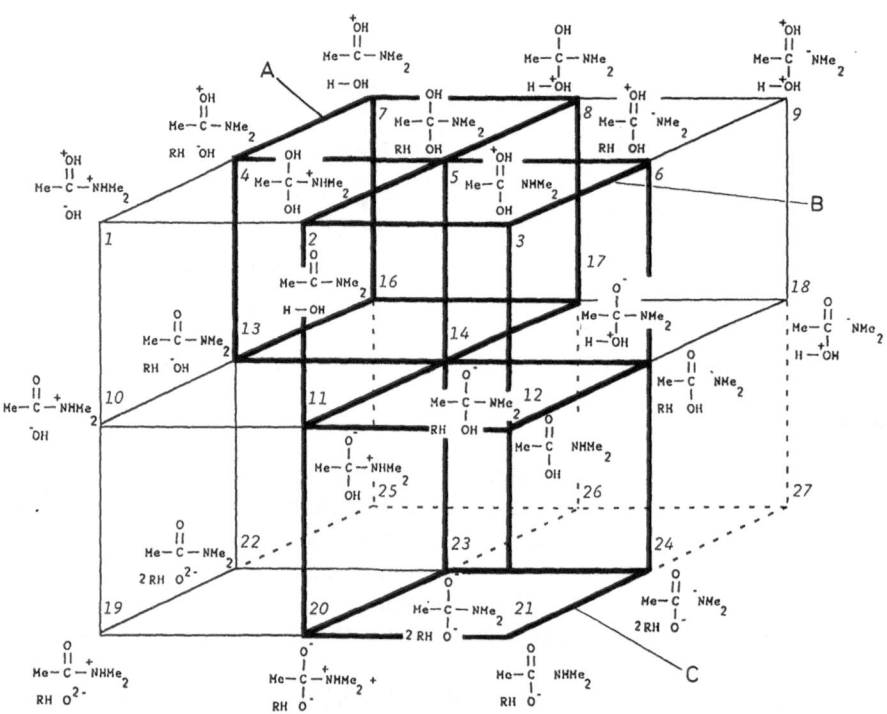

Fig. 15. Reaction cube for hydrolysis of N,N-dimethylacetamide

A cube A was obtained in the following way: system of N,N-dimethylacetamide and water (vertex 16) was solved in correspondence with work of Scudder [68], the same representation of processes by axes was accepted. Atoms $\diagdown C=$, $-\overline{O}-$, $=\overline{O}$, and $\diagdown \overline{N}-$ were successively provided as actual reaction centres. From generated synthons at the first level of the tree structures 5, 8, 13, 14, 16, and 17 were selected for cube A. At the first level synthons formed by changes of valence states at two atoms are not generated without at change

190

of their connecting edge, this is also true for synthons with double changed valence state on one atom. Therefore, structure *4* was not generated at the first level. For its generation, it is necessary to start from any structure of the neighboring vertex.

The cubes B and C were constructed from the first and second level synthons of synthesis tree respectively in a similar way. Relatively stable structures *5* and *12* were chosen as the educts for these cubes. Apart from synthons, which were preferably placed into the vertices of the reaction cubes, the program generated a set of other structures.

Structures of the type $\overset{\displaystyle R}{\underset{\displaystyle |\overline{O}^-}{\diagdown \diagup}}C^+ - \overline{N}R_2$ can be considered only as the extreme

cases (auxiliary formulas) on the way to stable states $\overset{\displaystyle R}{\underset{\displaystyle \overline{O}}{\diagdown \diagup\diagup}}C - \overline{N}R_2$. Some other

structures are isomeric systems

$$
\underset{\displaystyle \overset{|}{O^-}}{\overset{\displaystyle \overset{O}{||}}{- C -}} \quad
\underset{\displaystyle \overset{||}{O}}{\overset{\displaystyle \overset{O^-}{|}}{- C}}, \quad
\underset{\displaystyle H-O-H}{\overset{\displaystyle \overset{O}{||}}{-C-}} \quad
\underset{\displaystyle \overset{||}{O}}{\overset{\displaystyle \overset{O^- - *}{|}}{- C -}}
$$

which could be considered in reactions with labeled oxygen atoms.

Another series of generated structures

$$
\underset{\displaystyle H-* \, *-O-H}{\overset{\displaystyle \overset{O^-}{|}}{R- C = {}^+NR_2}}, \quad
\underset{\displaystyle H-O-H}{\overset{\displaystyle \overset{O- \, *}{|}}{R-C = {}^+NR_2}}, \quad
\underset{\displaystyle H-* \, O^- -H}{\overset{\displaystyle \overset{O^-}{|}}{R- C = {}^+NR_2}}
$$

anticipate states, which could take place on the face diagonal of eliminations.

It is necessary to say, that the PEGAS system produced other proposals corresponding to transformations of one of the educts, preferably an amide: except for a few less probable ketenes, imines, isocyanates etc., which are products of amide reactions under other reaction conditions, differences of the proposed conditions of hydrolysis can be mentioned.

After verifying the assumption that structures facing RC can be found in the first two levels of the synthesis tree where a comparison was made. For an adduct a structure was chosen common in the scheme to all completed RC *14*. From proposals generated in the first two levels of the tree of (retro)syntheses, it was possible to choose structures for all the corners of the neighboring cubes to correspond to the processes described by coordinate axes. All the vertices described in previous examples and in Ref. [68] were obtained. However, some others were also generated. The output is the scheme with six RC. Both of the dashed cubes need not be presented, because the vertices *25* to *27* would be equivalent to vertices *13* to *15* (the process described by moving along the Y axis is the opposite of the process described by the Z axis). The probability of the existence of the newly added structures is substantially low, but it is obvious that a reaction coordinate can also go via newly constituted cubes depending on reaction conditions.

The PEGAS system can also be used for obtaining proposals of possible structures (intermediates, charged tetrahedral intermediates) for the construction of RC. The choice of structures and their fitting into corners of the RC could easily be done automatically. In this way, the entire set of reaction cubes for individual possible reactions (in the synthetic and in the retrosynthetic direction) of the given set of educts or products (for the FVS considered) could be obtained. The user should then choose suitable cubes according to the processes placed on the axes of coordination as well as from the acceptability of structures placed in the corners RC.

5.3 Planning of Retrosynthesis

5.3.1 Methods of Forming a Benzthiazepinone Cycle Designed by PEGAS

Retrosyntheses of the benzthiazepinone, an analog of DTZ, has been also worked out by PEGAS.

	X	Y	configuration
I (DTZ)	S	$OCOCH_3$	2-S, 3-S
II	S	CH_3	cis
III	CH_2	$OCOCH_3$	cis

The target molecule considered by the program was the structure of the compound II without an alkyl substituent on the nitrogen atom. The alkylation of nitrogen is supposed to be the last step of syntheses. The PEGAS program generated the possible synthons regardless of reactivity and stereochemistry. Synthons, which appeared to be real from the reactivity aspects as evaluated by the chemist (localization of charge in the process of bond splitting, probability of supposed attack in the direct syntheses) were chosen from this wide offer. The individual routes suggested serve the chemist as inspiration for proposing alternative syntheses of target compound. The choosen paths were rewritten into the scheme of precursors and successors according to the included functional groups (Scheme 24).

When contemplating the assumed reactions in the direction of structures *2–6* in Scheme 24, the use of arylisocyanates could be considered as improper for

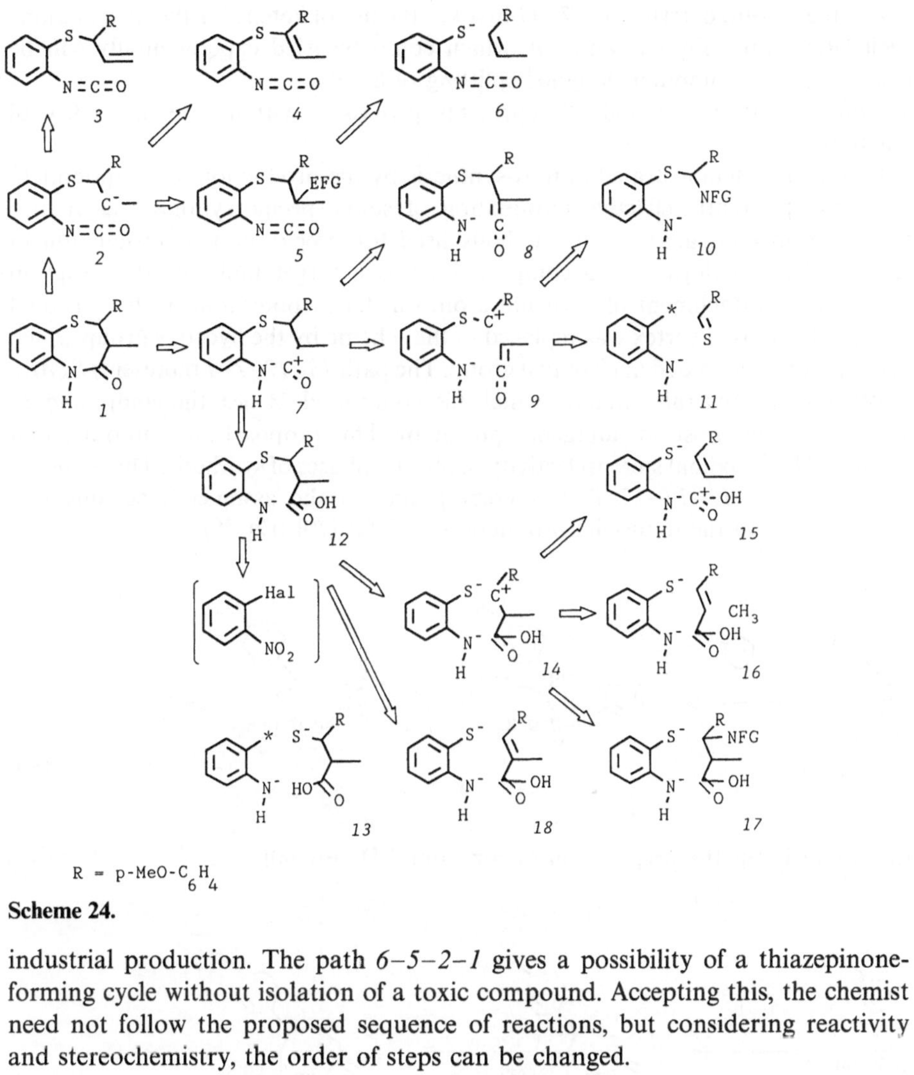

$R = p\text{-MeO-C}_6\text{H}_4$

Scheme 24.

industrial production. The path $6-5-2-1$ gives a possibility of a thiazepinone-forming cycle without isolation of a toxic compound. Accepting this, the chemist need not follow the proposed sequence of reactions, but considering reactivity and stereochemistry, the order of steps can be changed.

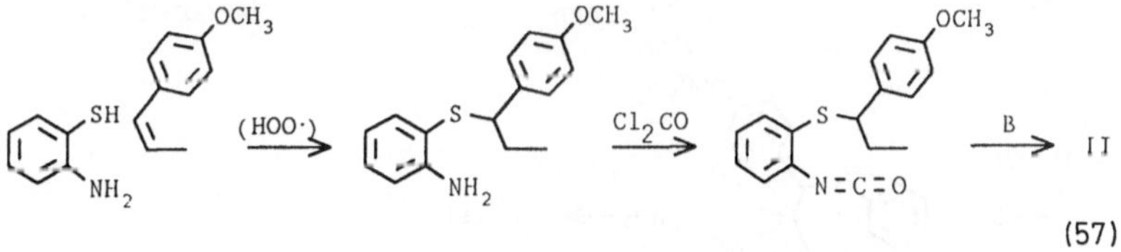

(57)

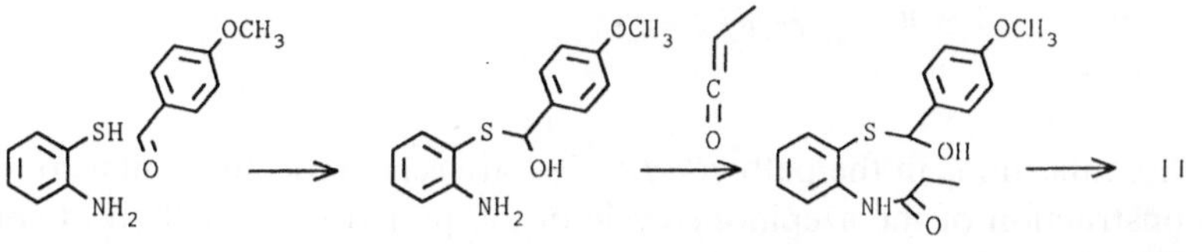

(58)

Another group of structures *7–11* invokes the use of ketene for the thiazepinone cycle. Reactions (Eqs. 57 and 58), which could be tried experimentally can be inspired by the computer designed splitting of bonds.

Note: Equations 57 and 58 represents proposals without stereospecifity of reactions.

Disjunction demonstrated in retrosynthesis by the production of compound *12* gives an apparently effective termination of stereospecific syntheses as it also follows from comparison with methods used for preparation or production of DTZ and its analogs. Using compound *12* as a target molecule the program proposed an attachment of a virtual atom on the carbon atom of the carbonyl group. This virtual vertex was replaced in the scheme by the hydroxy group which corresponds to a needed nucleofugal group. The path *13–12–7–1* represents $S_N Ar$.

When observing the paths on which the compound *14* lies, the compound *17* seems to be the most advantageous precursor. This proposal takes into account a demand for maximal stereoselectivity of the last phases of synthesis. The sequence of compounds *17–14–12–7–1* is corresponds to the assumed mechanism of reactions [69] in the industrial production of DTZ [70] (Eq. 59).

$$\text{(59)}$$

Alternatively for the preparation of compound II (Eq. 60).

$$\text{(60)}$$

Disconnections in the path *18–12–7–1* are schematically similar to those used in construction of the azepinone cycle in the preparation method of compound III (Eq. 61).

(61)

It is evident from the example presented that individual fragmentations of target molecules are successful offers to a chemist by the program. However, reactions are evoked regardless of reactivity and stereoselectivity. The chemist has to make the right choice using his/her experience for evaluation of these aspects. However, the chemist cannot be replaced by PEGAS or by any computer program, but the program can serve as a support for the chemist's imagination. Such a tool is useful because it never fails to put forward any possible reaction, even those seemingly inapplicable at first sight.

6 References

1. Balaban AT (ed) (1976) Chemical applications of graph theory. Academic, London
2. King RB (ed) (1983), Chemical applications of topology and graph theory. Elsevier, Amsterdam
3. Lindsay RK, Buchanan BG, Feigenbaum EA, Lederberg J (1980) Applications of artificial intelligence for organic chemistry: The DENDRAL Project. McGraw-Hill, New York
4. Pierce TH, Hohne BA (eds) (1986), Artificial intelligence application in chemistry. ACS, Washington, DC
5. Vleduts GE (1963) Inf Storage Ret, 1: 101
6. Corey EJ, Wipke WT (1969) Science 166: 178
7. Vernin G, Chanon M, editors (1986), Computer aids to chemistry. Ellis Horwood, Chichester. Cf also serie of papers in the recent issue of the Anal Chem Acta (1990) 235(1): 1
8. Wipke WT, Rogers D (1984) J Chem Inf Comput Sci 24: 71
9. Bures MG, Jørgensen WL (1988) J Org Chem 53: 2504
10. Hendrickson JB (1976) In: Boschke FL (ed) Synthetic and mechanistic organic chemistry, Topics in Current Chemistry, vol 62, p 49. Springer, Berlin Heidelberg New York
11. Hendrickson JB (1990) Anal Chim Acta 235: 103
12. Moreau G (1978) Nouv J Chim 2: 187
13. Dugundji J, Ugi I (1973) Topics Curr Chem 39: 19
14. Doenges R, Groebel BT, Nickelsen H, Sander J (1985) J Chem Inf Comput Sci 25: 425
15. Zefirov NS, Gordeeva EV (1987) Usp Chim LVI(10): 1753
16. Koča J, Kratochvíl M, Kvasnička V, Matyska L, Pospíchal J (1989) Synthon model of organic chemistry and synthesis design, Lecture Notes in Chemistry, vol 51 Springer, Berlin Heidelberg New York

17. Matyska L, Koča J (1991) J Chem Inf Comput Sci 31: 380
18. Koča J (1988) Coll Czech Chem Comm 53: 1007
19. Koča J (1988) Coll Czech Chem Comm 53: 3108
20. Koča J (1988) Coll Czech Chem Comm 53: 3119
21. Koča J (1989) Coll Czech Chem Comm 3: 73
22. Koča J (1989) J Math Chem 3: 91
23. Corey EJ (1967) Pure Appl Chem 14: 19
24. Kvasnička V, Pospíchal J (1990) Int J Quant Chem XXXVIII: 253
25. Kvasnička V, Pospíchal J (1989) J Math Chem 3: 161
26. Harary F, Palmer ME (1973), Graphical Enumeration. Academic Press, New York
27. Ugi I, Bauer J, Brandt J, Friedrich J, Gasteiger J, Jochum C, Schubert W (1979) Angewandte Chemie 2: 99
28. Koča J, Kratochvíl M, Kunz M, Kvasnička V (1984) Coll Czech Chem Comm 49: 1247
29. Kvasnička V, Kratochvíl M, Koča J (1983) Coll Czech Chem Comm 48: 2284
30. Baláž V, Kvasnička V, Pospíchal J (1992) Discr Appl Math 35: 1
31. Ingold KC (1969) Structure and mechanisms in organic chemistry. Cornell University Press, Ithaca
32. Balaban AT, Motoc I, Bonchev D, Mekenyan O (1983) In: Boschke FL (ed) Steric effects in drug design, p 21. Akademie-Verlag, Berlin
33. Stankevich MI, Stankevich IV, Zefirov NS (1988) Russian Chemical Reviews 57: 191
34. Clerc JT, Terkovics AL (1990) Anal Chim Acta 235: 93
35. Pauling L (1931) J Amer Chem Soc 53: 1367
36. Vleck JHV (1934) J Chem Phys 2: 20
37. Kratochvíl M (1983) Chem Listy 77: 225
38. Weise A (Personal Communication)
39. Kratochvíl M, Koča J, Kvasnička V (1984) Chem Listy 78: 1
40. Wochner M, Brandt J, von Scholley A, Ugi I (1988) Chimia 42(6): 217
41. Mezey PG (1988) J Math Chem 2: 299
42. Johnson MA, Gifford E, Tsai CC (1990) In Johnson MA, Maggiora GM, editors, Concepts and applications of molecular similarity. John Wiley, New York
43. Arteca GA (1990) Theor Chim Acta 78: 377
44. Kvasnička V, Pospíchal J, Koča J, Hladká E, Matyska L (1990) Chem Listy 84: 449
45. Koča J, Kratochvíl M, Matyska L, Kvasnička V (1986) Coll Czech Chem Comm 51: 2637
46. Kratochvíl M (1985) Chem Listy 79: 807
47. Kvasnička V, Kratochvíl M, Koča J (1987) Mathematical chemistry and the computer aided organic synthesis. Academia, Praha (in Czech)
48. Matyska L, Jůhová K, Mikula P, Potůček V, Kratochvíl M (1990), The Program Package PEGAS: I. Manual, II. Annotated Examples, Organic synthesis and computers vol 9. Dům techniky ČS VTS, Pardubice
49. Shelley CA (1983) J Chem Inf Comput Sci 23: 61
50. Matyska L, Koča J (1991). J Chem Inf Comput Sci 31: 380
51. Schubert W, Ugi I (1978) J Am Chem Soc 100: 37
52. Munk ME, Lind RJ (1986) Anal Chim Acta 184: 1
53. Moffat JB (1979) J Mol Struct 52: 275
54. Granier M, Baceiredo A, Dartiguenave Y, Dartiguenave M, Menu M, Bertrand G (1990) J Am Chem Soc 112: 6277
55. Schonberg A, Moubacher R (1952) Chem Rev 50: 261
56. Vaidyanathan G, Wilson JW (1989) J Org Chem 54: 1815
57. Gröbel BT, Seebach D (1977) Chem Ber 110: 852
58. Hudrlik PF, Agwarambgo ELO, Hudrlik AM (1989) J Org Chem 54: 5613
59. Vedejs E, Meier GP (1983) Angew Chem, Int Ed Engl 22: 56
60. Schneider WP (1969) J Chem Soc, Chem Commun p 785
61. Olah GA, Krishnamurthy VV (1982) J Am Chem Soc 104: 3987
62. Vedejs E, Marth CF (1990) J Am Chem Soc 112: 3905

63. McKee ML, Paul GP, Shevlin PB (1990) J Am Chem Soc 112: 3374
64. Meier IK, Schwartz J (1989) J Org Chem 111: 3069
65. Matyska L, Hladká E, Pospíchal J, Koča J, Kratochvíl M, Kvasnička V (1989) Chemistry in Graphs III. Lachema Brno (in Czech)
66. Bates RB, Ogle CA (1983) In: Topics in Current Chemistry, vol 114, p 70 Springer, Berlin Heidelberg New York
67. Zhang JJ, Schuster GB (1988) J Org Chem 53: 716
68. Scudder PH (1990) J Org Chem 55: 4238
69. Kugita H, Inone H, Ikezaki M, Takeo S (1970) Chem Pharm Bull 18: 2028
70. Běluša J, et al (1991) CS Patent 274213
71. Pospíchal J, Kvasnička V (1990) Theor Chim Acta 76: 423
72. Kvasnička V, Pospíchal J (1990) J Math Chem 5: 309

New Elements in the Representation of the Logical Structure of Chemistry by Qualitative Mathematical Models and Corresponding Data Structures

Ivar Ugi, Natalie Stein, Michael Knauer, Bernhard Gruber, Klemens Bley

Organisch-Chemisches Institut, Technische Universität München, Lichtenbergstr. 4, 8046 Garching, FRG

Rupert Weidinger

Institut für Informatik, Universität Passau

Table of Contents

Topics in Current Chemistry, Vol. 166
© Springer-Verlag Berlin Heidelberg 1993

1 Introductory Remarks

1.1 Mathematics and Theories in Chemistry

Mathematics is omnipresent in physics, and in physics all theories have a strong mathematical aspect. The collection and evaluation of experimental data of well-defined systems belong to the prevalent activities of physicists, who generally require that their theories are suitable for the quantitative interpretation and prediction of the numerical results of observations.

Mathematics plays a less dominant role in chemistry. Mostly mathematics has entered chemistry via physics. Physics-based quantum chemistry is so widely accepted as a theoretical device in chemistry that the term quantum chemistry is almost a synonym for theoretical chemistry. In general, quantum chemistry is used to generate numbers that can be compared with numerical values of experimental results concerning discrete molecular objects. Like the other physics-oriented theories in chemistry, quantum chemistry is not a true chemical theory, since it can not be used for the direct solution of chemical problems, but it provides physical data of molecules that may be used indirectly to interpret and to predict the chemical properties and behaviour of molecular systems.

In contrast, truly chemical problems do not have quantitative numerical solutions. The unknown "X" of a chemical problem is usually a molecular system or a chemical reaction. The most important parts of chemical knowledge are qualitative in nature, and in chemistry numerical data is often only important, because it provides qualitative insights.

After the conceptual edifice of modern theoretical physics had been erected, Dirac [1] pronounced the contemporary spirit of physics: "The underlying physical laws necessary for the mathematical theory of a large part of physics and the whole of chemistry are thus completely known, and the difficulty is only that the exact application of these laws leads to equations much too complicated to be soluble."

Dirac understood the fact that the majority of chemical systems and phenomena are not representable by idealized models, that many chemical problems involve complex relations to other, often unknown, chemical systems, and also probably that the non-deterministic and combinatorial aspects of chemistry are insurmountable obstacles to the solution of many chemical problems by explicit computation.

Also, when using the methods of theoretical physics, one must know the right tree to bark up. However, in chemistry, one often does not even known the forest where that tree is.

The physics-based mathematical theories of chemistry are clearly very useful and often indispensable, but at the same time, their limitations are well-known.

Can mathematics be used for the solution of genuine and typical chemical problems? Is the mathematization of chemistry possible without the intermediacy of physics and beyond the customary approaches?

Since the physics-mediated use of quantitative mathematics in chemistry is nearing a plateau, progress in the solution of chemical problems by numerical computation can hardly be expected to lead to the required fundamentally new theoretical approaches to chemistry.

Although qualitative mathematical concepts, models and theories have been successfully used in chemistry for more than 100 years [2], new territory in this area is by far not yet exhausted.

Sir Arthur Cayley [3] introduced the application of graph theory to chemistry. Meanwhile chemical graph theory has developed into an important and rapidly growing branch of mathematical chemistry [4].

Through essentially group theoretical reasoning van't Hoff and Le Bel were led to postulate the asymmetric carbon atom [5]. This was the beginning of stereochemistry. Other spectacular applications of group theory are the enumeration of isomers by Polyá [6], De Bruijn [7], Ruch et al. [8a], and the group theoretical approaches to chirality [8b].

From 1970 to 1984 Dugundji, Ugi et al. [9, 10] analyzed the logical structure of chemistry. They found that chemistry consists of two parts with distinct logical structures, constitutional chemistry and stereochemistry. This division is already implied by the nongeometric definition of stereoisomers [11] (see Sect. 1.3). Accordingly, Dugundji and Ugi formulated a qualitative mathematical model of the logical structure of chemistry that consists of two distinct parts.

The theory of the *be-* and *r*-matrices [9] represents the logical structure of constitutional chemistry.

The theory of chemical identity groups [10] is based on the concept of permutational isomerism [12] and represents the logical structure of stereochemistry.

1.2 From Computational Chemistry to Computer Chemistry

As soon as computers became available, they were used in chemistry. As the various hardware/software capabilities of computers advanced, their exploitation in chemistry followed suit.

The oldest and most widespread applications of computers in chemistry rely on the *numerical computing power* of computers. This discipline is called *computational chemistry*. Nowadays numerical computation in absolutely indispensable in many areas of chemistry and for many methods of chemistry, such as quantum chemistry, as well as the collection and evaluation of experimental data, e.g. in X-ray crystallography and FT-NMR.

Modern chemical documentation is inconceivable without the *data-storage/retrieval capabilities* of computers.

The importance of the *graphic capabilities* of computers is rapidly increasing. More and more computer applications in chemistry rely on graphic input/output systems, and the popularity of so-called modelling systems is growing at an astounding rate.

The aforementioned types of computer applications in chemistry have strongly entered the routine work of chemistry, whereas the computer programs that are based on the logical capabilities of computers have not yet been accepted equally well by their chemical users. Since the 1960s various synthesis design programs and other computer programs for the solution of chemical problems have been developed [13 – 15]. Some of these have reached high levels of performance, but in chemistry they are still used much less than they deserve.

Although this field, so-called *computer chemistry*, is more than 20 years old, there is still a gold rush spirit, and good mining habits have not yet been fully established.

There are two profoundly different approaches to computer chemistry, the information-oriented approach, and the logic-oriented approach. Corey, Wipke, Gelernter et al. [13, 14] have developed diverse retrosynthetic synthesis design programs that are based on so-called reaction libraries [16]. Here problems are primarily solved by retrieval and manipulation of stored empirical data. This has the advantage that all the solutions produced are chemically realistic, since they are directly derived from some of the stored chemical information. On the other hand, such programs are unable to generate any chemistry beyond an analogy to the known.

Computer programs for chemistry may also be based on a suitable theory, without relying on any stored detailed empirical data. Ugi et al. [17–19] have developed diverse computer programs for the deductive solution of chemical problems on the basis of the theory of the *be-* and *r*-matrices. The term "deductive" refers to the fact that the conceivable solutions to a particular problem are deduced from general principles that have been incorporated into the theory of the *be-* and *r*-matrices. The disadvantage of this method is that, in general, for any given chemical problem an immense number of solutions is generated by exhaustive procedures. Many of these solutions are realistic, or even belong to known chemistry, but there may also be very many solutions that are not at all close to known chemical reality. The advantage of the theory-based programs is, that novel types of results can be obtained. Recently some new chemistry has been discovered by the use of the theory-based computer programs IGOR2 [18–21] and RAIN2 [22–24].

Since very many solutions to a given chemical problem may be produced by such computer programs, the few desirable solutions must be picked from the sometimes large number of potential solutions that have been generated.

Some of the chemical computer programs use heuristic rules to select the desirable solutions [13, 14], others rely on estimates of physico-chemical data, e.g. reaction enthalpies, electronegativities, charge affinities or polarizabilities [25] as a basis for selection procedures, which is also heuristic in nature. As an alternative to the heuristically oriented selection procedures, the solutions to a chemical problem can also be classified and selected by formal means [19, 20, 26]. Since, as a rule, formal selection procedures still yield more solutions than the few acceptable ones, such a selection procedure must be combined with an interactive selection by the user. This has the advantage, that a transparent and non-arbitrary selection process results in which the capabilities of the computer and the chemical knowledge, expertise, imagination and intuition of the user are fully exploited.

It is not clear which of the competing approaches to generating and selecting the solutions to chemical problems will ultimately be more acceptable to the chemical usership. The users will decide which type of computer-assistance to chemistry will prevail in the end. This also involves a competition of two philosophies. One is the philosophy of some engineers, who are happy, "when it works", without caring why it works. The other is the philosophy of the true scientist, who is also happy when it works, but who, at the same time, wants it to work for transparent and scientifically justified reasons.

1.3 Isomerism and the Logical Structure of Chemistry

Since 1970 [12] the term "logical structure of chemistry" has been used for a system of relations between the objects of chemistry. Isomerism, the interconvertibility by chemical reactions and the other types of chemical similarity [19] belong to the most important relations between molecular systems. The logical structure of chemistry is determined by the valence chemical properties of the approximately 100 chemical elements [27].

Any two distinguishable molecules are said to be isomeric, if they have the same empirical formula and are based on the same collection of atoms. The hierarchic classification of molecules, according to isomerism, constitutional isomerism and stereoisomerism, is a centerpiece of the logical structure of chemistry [28].

The emergence of chemistry as a science and the evolution of many basic concepts in chemistry are closely tied to the discovery of isomerism by Berzelius, Liebig, Wöhler and its further investigation by other classical chemists in the early 19th century [2] and A. v. Humboldt's insight [29] that the existence of isomerism implies an internal structure of molecules, the smallest units of pure chemical compounds. The formulation and general acceptance of the structure theory in 1860 [2], the beginning of a new area of organic chemistry, as well as the development of stereochemistry since 1874 [5], are inseparable from the progress in the conceptual and experimental study of isomerism.

An extension of isomerism from molecules to ensembles of molecules (EM) leads to new perspectives in chemistry. The left and right hand sides of a stoichiometrically balanced reaction equation are isomeric EM. Any chemical reaction may be regarded as an isomerization, i.e. the conversion of an EM into an isomeric EM. Let $A = \{A_1, \ldots A_n\}$ be a finite collection of atoms with the empirical formula A. Any EM that contains each atom of A exactly once is an EM(A). The family of all isomeric EM(A), the FIEM(A) contains the complete chemistry of $A = \{A_1, \ldots A_n\}$. The FIEM(A) is closed and finite. It has well-defined limitations and invariancies. Accordingly, the logical structure of the chemistry of an FIEM(A) is much easier to elucidate than the logical structure of chemistry without the above restrictions [9].

In "structure theory" the term "structure" refers to what is now called the chemical constitution of a molecule. Now "structure" refers to the three-dimensional structure of a molecule, as determined by X-ray crystallography. The chemical constitution of a molecule is described by stating for each atom its covalent bonds and the neighbouring atoms to which it is connected by these covalent bonds. The chemical constitution of a molecule is visualized by its constitutional formula, a representation of the molecule by a graph whose vertices are labelled by chemical element symbols representing the cores (nuclei and inner shell electrons) of the respective atoms. The edges are labelled by the formal orders of the covalent bonds.

A covalent bond of the formal order one corresponds to a pair of valence electrons that are shared by two atomic cores that are thus connected. The valence electrons that are not engaged in bond forming are called lone valence electrons.

The chemical constitution of a molecule can also be described by so-called connectivity matrices [30] whose rows/columns are assigned to the atomic cores. An off-diagonal entry of a connectivity matrix is the formal bond order of the bond between the two atoms represented by the row and column that intersect at the given entry. The diagonal of a connectivity matrix may be used to label the rows/columns by chemical element symbols.

A *be*-matrix is obtained from a connectivity matrix by replacing the chemical element symbols in the diagonal by the numbers of lone valence electrons of the atoms [9].

The chemical constitution of a molecule or EM with integer covalent bond orders is representable by a *be*-matrix $B = \langle b_{i,j} \rangle$. For an EM(A) with n atoms its *be*-matrix B is an n × n symmetric matrix with positive integer entries $b_{i,j} = b_{j,i}$. An off-diagonal entry $b_{i,j}$ is the formal covalent bond order between the atoms A_i and A_j. The diagonal entry $b_{i,i}$ is the number of lone valence electrons at the atom A_i.

The *be*-matrices are particularly well-suited for computer-assisted manipulation of constitutional information. In fact, the *be*-matrices are true mathematical objects, and through *be*-matrices chemical problems can be solved by solving the corresponding mathematical problems.

Isomers that differ by their chemical constitution are constitutionally isomeric. As traditionally defined, stereoisomers are molecules with the same chemical constitution that differ with respect to the relative spatial arrangement of their constituent atoms. Since many types of flexible molecules exist whose shape rapidly change with time and that are not adequately representable by any geometric model, stereoisomers must be defined as follows, without reference to molecular geometry:

Any two molecules are stereoisomers, if they have the same chemical constitution, in common but are not chemically identical. □

Molecules are chemically identical if they undergo spontaneous interconversion under given observation conditions and belong to the same pure chemical compound. If molecules are not chemically identical, they are chemically distinct, and belong to different chemical compounds.

According to these definitions of constitutional chemistry and stereochemistry, they are two disjoint aspects of chemistry. It follows that their treatment by qualitative mathematics must differ profoundly. The logical structure of constitutional chemistry is represented by the algebra of *be*- and *r*-matrices. The logical structure of stereochemistry is essentially permutation group theoretical in nature.

2 Mathematical Models of the Logical Structure of Constitutional Chemistry

The chemistry of an FIEM(A) of a finite collection of atoms is a finite closed system with a variety of well-defined properties. Thus its logical structure is easier to analyze than the logical structure of chemistry in general. Since any collection of atoms may serve as A, any qualitative relations that exist in an FIEM(A) are analogously also valid for chemistry as a whole [9].

Thus, the investigation of the constitutional aspect of the chemistry of an FIEM(A) provides us with knowledge about the constitutional aspect of the logical structure of chemistry.

The representation of an $EM_B(A)$ by its *be*-matrix $B = \langle b_{i,j} \rangle$, an n × n symmetric matrix with positive integer entries and well-defined algebraic properties, corresponds to the act of translating an object of chemistry into a genuine mathematical object. This implies that the chemistry of an $EM_B(A)$ corresponds to the algebraic properties of the matrix B.

A row/column of a *be*-matrix represents the valence scheme of an atom of the chemical element to which it is assigned. Accordingly, the rows/columns of *be*-matrices must correspond to the allowable valence schemes of the respective chemical elements. The sum over the entries of a row/column in a *be*-matrix is the number of valence electrons that belong to the respective atom. The sum of the entries in a *be*-matrix is the total number of valence electrons in the EM.

Note that, being a labelled graph, the constitutional formula of an EM is also a mathematical object. Its algebraic counterpart, the *be*-matrix, is suitable for mathematical manipulations by a computer, while the constitutional formula is more convenient for a chemist who prefers visual information.

Any two distinct *be*-matrices B and B′ represent the same $EM_B(A)$ if they are interconvertible by row/column permutations that corresponds to permutations of the atomic indices in A. Such a row/column permutation is achieved by the action of a corresponding permutation matrix P

$$B' = PBP^{-1}.$$

Accordingly an $EM_B(A)$ is representable by up to n! distinct but equivalent *be*-matrices [31]. One of these *be*-matrices of the $EM_B(A)$ may be chosen as the canonical representation, e.g. by the algorithm CANON [32]. Note that any constitutional symmetries in $EM_B(A)$ are also detected by CANON.

The conversion of an $EM_B(A)$ at the Beginning of a reaction and an $EM_E(A)$ at the End of a reaction

$$EM_B(A) \rightarrow EM_E(A)$$

corresponds to a redistribution of valence electrons while the atomic cores of A remain unchanged. This reaction is represented by a transformation B → E of the *be*-matrix B into the *be*-matrix E. This transformation is achieved by elementwise addition of the *r*-matrix (Reaction matrix) $R = E - B$, according to

$$B + R = E.$$

The *r*-matrix $R = \langle r_{i,j} \rangle$ is an $n \times n$ symmetric matrix. Its off-diagonal entries $r_{i,j}$ $(i \neq j)$ indicate the changes in formal bond order between A_i and A_j, and the diagonal entries $r_{i,i}$ represent changes in the number of lone valence electrons at A_i. The sum of the entries $r_{i,j}$ is

$$\sum_{1 \leq i,j \leq n} r_{i,j} = 0,$$

while the sum

$$D(B, E) = \sum_{1 \leq i,j \leq n} |r_{i,j}| > 0$$

of the absolute values of the entries is twice the number of valence electrons that are redistributed during the reaction $EM_B(A) \rightarrow EM_E(A)$.

We have $D(B, B) = 0$, $D(B, E) = D(E, B)$, and $D(B, X) + D(X, E) \geq D(B, E)$, i.e. $D(B, E)$ is a distance in a mathematical sense. We call $D(B, E)$ the *chemical distance* (CD) between $EM_B(A)$ and $EM_E(A)$.

A geometric interpretation follows from mapping the $n \times n$ *be*-matrices B onto points P(B) in an n^2-dimensional euclidean space. There $D(B, E)$ is the L_1-distance, the "city block"-or "taxi driver"-distance, between P(B) and P(E). Thus an FIEM(A) corresponds to a lattice of points in an n^2-dimensional euclidean space. The reaction matrices correspond to vectors between the *be*-points [9, 19, 33].

Since an EM is representable by up to n! distinct *be*-matrices, it corresponds to a cluster of up to n! *be*-points. Unless specified otherwise, the term CD between $EM_B(A)$ and $EM_E(A)$ is used for the CD between the "closest" *be*-points of $EM_B(A)$ and $EM_E(A)$.

In the chemical applications of this model the problem and its solutions are imbedded into a suitable FIEM. For instance, in bilateral synthesis design the $EM_B(A)$ of the starting materials and the target $EM_E(A)$, i.e. the target molecule and its coproducts, correspond to the *be*-points P(B) and P(E), and the pathways that connect P(B) with P(E) via the *be*-points of intermediate EMs are the conceivable syntheses. The solutions of such chemical problems are found by solving the fundamental equation

$$B + R = E$$

of the model [34]. When B represents the collection of starting materials of a synthesis, the next level of intermediates is found through the pairs (R, E) that satisfy the fundamental equation under given suitable boundary conditions. The geometry of the n^2-dimensional representation space can be used to selectively generate those intermediates that lie within a suitably chosen envelope around P(B) and P(E) [19, 26] and thus belong to a relatively "short" synthetic pathway.

3 The Interpretation of Stereochemistry by Permutational Isomerism and the Theory of Chemical Identity Groups

Stereochemistry is the discipline of chemistry that relates the observable chemical properties and behaviour of chemical compounds to the geometric features of their molecules. Stereochemistry is mostly concerned with stereoisomerism and its observable consequences.

Elementary geometry and the associated theory of point group symmetry are still the main mathematical tools of stereochemistry, despite the increasing importance of flexible molecules that are not adequately representable by rigid geometric models [10, 11, 35].

A unified general interpretation and formal treatment of current stereochemistry cannot be based on geometry. A more abstract point of view is needed. Chemical identity and permutational isomerism are notions that are suitable as conceptual foundations for a unified theoretical approach to stereochemistry.

In 1936 Polyá [6] introduced a method for the enumeration of isomers that is based on the placement of ligands at skeletal sites. Subsequently, the classification and enumeration of isomers became an active field of investigations [7, 8].

Permutational isomerism was defined in 1970 [12]. It relies on a conceptual dissection of molecules into a set of ligands L and a skeleton in a fashion that is appropriate for a considered problem [10]. Any distinct molecules that result from a given molecule by permuting its ligands are called permutationally isomeric. The set of all isomers that can be generated in this way is called a *family of permutation isomers*. Although stereoisomers exist that are not permutation isomers, and there are non-steroisomeric permutational isomers, the notion of permutational isomerism in combination with the concept of chemical identity serves well as a basis for a unified essentially non-geometric treatment of stereochemistry [10, 35].

The molecules of a pure and uniform compound X are chemically identical, and yet they can differ greatly in shape. We select one molecular individual from the compound X as characteristic model E, and we define its set of ligands and its skeleton in terms of a set of ligand sites. We assume that E has n chemically distinguishable ligands.

Each permutation of the ligands of E gives a molecule representing some chemical compound, not necessarily X. The *chemical identity group* S_X of X consists of those permutations of the ligands of E that lead to a molecule belonging to X. Note that the ligand permutations that preserve chemical identity do not represent symmetry operations. They do not even necessarily bring the skeleton into self-coincidence, as is required by conventional representations of symmetries by skeletal site permutations. The set of all permutations of the ligands in L forms a group SymL, the symmetric group of $|L|$ objects. The chemical identity group S_X of X is a subgroup of SymL. The chemical identity group is the conceptual foundation of the present approach to the stereochemistry of molecules, flexible or not. If all ligands of X are chemically distinct, the left cosets λS_X of S_X in SymL correspond to the distinct permutation isomers of X, i.e. X has

$$|SymL|/|S_X| = n!/|S_X|$$

permutation isomers. These left cosets or any of their elements can be used as permutational descriptors in a permutational nomenclature.

Let E be the reference model of a pentacoordinate molecule with a D_{3h} skeleton (see below). Then (123) E, (132) E, (12)(45) E and (23)(45) E are rotated forms of E, and they all represent the same permutation isomer X.

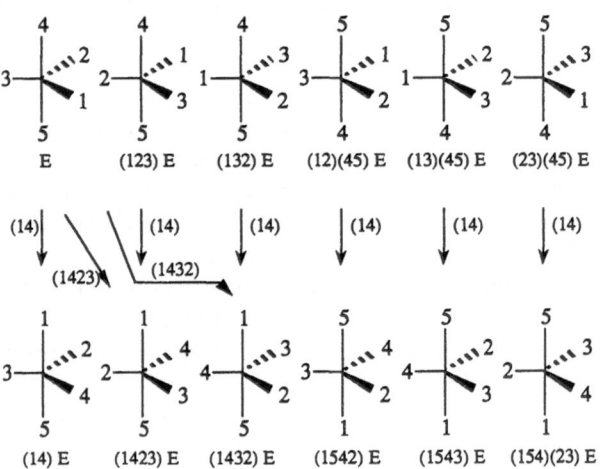

Fig. 1. Ligand permutations

Thus the chemical identity group is

$$S_X = \{e, (123), (132), (12)(45), (13)(45), (23)(45)\} .$$

The model E is converted into the model (14) E by the ligand permutation (14). The action of (14) on the other models of X, as represented by the elements of S_X, yields the models of the left coset (14) S_X. The models of the isomer (14) X are obtained by the action of a ligand permutation (belonging to S_X, followed by the ligand permutation (14). These all belong to the permutation isomer (14) X, and they all correspond to this isomer

$$(14) S_X = \{(14), (1423), (1432), (1542), (1543), (154)(23)\} .$$

In the theory of *be-* and *r*-matrices the constitutional aspect of an EM is represented by the *be*-matrices, and the chemical reactions are represented by additive transformations of the *be*-matrices by the action of the *r*-matrices.

Correspondingly, in the theory of the chemical identity groups stereochemical features of the molecules and EMs are represented by their chemical identity groups and their left cosets in SymL, and the permutational isomerizations and the stereochemical aspect of chemical reactions are described by the so-called set-valued mappings of the left coset spaces of the respective chemical identity groups [10, 19, 35].

Let X and Y be isomers with the same set of ligands, and let the isomerization X → Y be a reference process, then processes with the same mechanism will convert X into those permutation isomers of Y that are represented by left cosets μS_Y with non-empty intersections $X \cap \mu Y \neq \varnothing$, and μY will also be analogously convertible into the permutation isomers λX, if we have $\lambda X \cap \mu Y \neq \varnothing$.

We return to the above example X → (14) X ≡ Y. The chemical identity group of (14) X is:

$$S_Y = S_{(14)X} = (14) S_X(14) = \{e, (234), (243), (15)(23), (15)(24),$$
$$(15)(34)\} .$$

The chemical identity group S_X intersects with the cosets μS_y if $\mu \in S_X$, i.e. $S_X \cdot S_Y$ contains the left cosets $\mu S_Y (\subset S_X \cdot S_Y)$ representing the permutation isomers of Y that are reached by processes in analogy to X → Y. For instance X will also be converted into $(12)(45) Y((12)(45) \in S_X)$ that is represented by:

$$(12)(45) S_Y = \{(12)(45), (13452), (14532), (13254), (14)(25), (12534)\} .$$

Similarly, $S_Y \cdot S_X$ indicates the isomers that are obtained from Y.

When chemical reactions with a constitutional aspect and a stereochemical aspect are written as ligand preserving reactions, then their stereochemical aspect is accounted for by set-valued mappings of the left cosets of the chemical identity groups of the EM of educts and the EM of products. Thus, for example, one can immediately tell how many and which stereoisomeric products may conceivably result from a given reaction [35],

The above formalism applies to permutation isomers whose ligands are all chemically distinct. If, however, some of the ligands are not chemically distinguishable, this must be taken into account by representing the isomers by double cosets $\Sigma \lambda S_X$ of the chemical identity group S_X and the so-called stabilizer groups Σ that stand for the equivalencies of ligands. The double cosets $\Sigma \lambda S_X$ in SymL are obtained by a set-valued mapping of the right cosets $\Sigma \lambda$ of Σ in SymL and the left cosets λS_X of S_X in SymL [10].

4 Overcoming the Limitations of *be-* and *r*-Matrices as a Basis for Computer Assistance in Chemistry

The algebra of *be-* and *r*-matrices is well-suited as a theoretical basis for computer programs that solve problems of constitutional chemistry, provided that the formal covalent bond orders are representable by integers. With molecular systems having delocalized electrons or multicenter bonds the representation of EM by *be*-matrices is not adequate, because fractional bond orders are encountered. In particular, the translation from *be*-matrices into constitutional formulas meets with difficulties.

A valid alternative for systems of delocalized electrons is to represent the considered molecules by their canonical forms according to VB-theory. However, this approach does not reflect chemical reality and it is certainly not adequate for molecules with multicenter bonds and many organometallic systems.

In order to avoid the aforementioned difficulties we propose an extension of the theory of *be*- and *r*-matrices. This model is closely related to the VB-theory. The extended theory of the *be*- and *r*-matrices contains the latter as a subset, and it reflects an MO-approach.

The theory of the extended *be*-matrices (*xbe*-matrices) and the extended *r*-matrices (*xr*-matrices) is a representation of a constitutional chemistry which includes EM with systems with delocalized valence electrons and multicenter bond systems [36]. We shall use the collective term DE-systems for these. We consider the $EM_B(A)$, a molecule or an EM that is based on a collection $A = \{A_1, ..., A_n\}$ of n atoms. If $EM_B(A)$ does not contain DE-systems, it is adequately described by its $n \times n$ *be*-matrix B (see section 2). If, however, $EM_B(A)$ contains m DE-systems, and the k-th DE-system comprises α_k atomic cores and β_k valence electrons, then the localized covalent bonds and lone valence electrons are given by the customary $n \times n$ *be*-matrix B of $EM_B(A)$ where the delocalized electrons are neglected, and the corresponding *xbe*-matrix XB is obtained from B through augmenting B by m rows/columns, one for each DE-system. The diagonal entry $b_{n+k,n+k}$ of the n + k-th row/column is the number of valence electrons β_k that belong to the k-th DE-system. An off-diagonal entry $b_{i,n+k} = b_{n+k,i}$ ($1 \leqq i \leqq n$; $1 \leqq k \leqq m$) is 0 if the atom A_i does not belong to the k-th DE-system, and is 1 if A_i is part of the k-th DE-system.

Since the *xr*-matrices are differences of *xbe*-matrices, they have the following properties: An *xr*-matric XR is a symmetric $n \times n$ matrix whose non-zero entries are positive or negative integers. The entries $r_{i,j}$ ($1 \leqq i, j \leqq n$) are the same as in the *r*-matrices. The diagonal entries $r_{n+k,n+k}$ indicate the change in the number of valence electrons that belong to the k-th DE-system. The off-diagonal entries $r_{i,n+k} = r_{n+k,i}$ indicate whether the atom A_i loses ($r_{i,n+k} = -1$) or gains ($r_{i,n+k} = +1$) membership in the k-th DE-system. $r_{i,n+k} = 0$, if A_i is not involved in the k-th DE-system.

For *xr*-matrices we have

$$\sum_{1 \leqq i, j \leqq n} r_{i,j} + \sum_{1 \leqq k \leqq m} r_{n+k,n+k} = 0$$

while

$$d(XB, XE) = \sum_{1 \leqq i, j \leqq n} |r_{i,j}| + \sum_{1 \leqq k \leqq m} |r_{n+k,n+k}| \geqq 0$$

is the CD between $EM_B(A)$ and $EM_E(A)$, two distinct isomeric EM.

When a DE-system of μ valence electrons and ν atomic cores is converted by a chemical reaction into a system of localized covalent bonds and lone valence electrons, the new localized covalent bonds and lone electrons must involve the atomic cores that belonged to the DE-system, and at least all but one of the newly formed covalent bonds must be between atoms of the previous DE-system. The reverse holds for reactions that convert localized bonds and electrons into DE-systems. This corresponds to boundary conditions for the placement of the respective entries $r_{i,j} = r_{j,i} \neq 0$ $(1 \leqq i, j \leqq n)$ in *xr*-matrices.

A representation of the Wacker Reaction [37] may serve as an illustration:

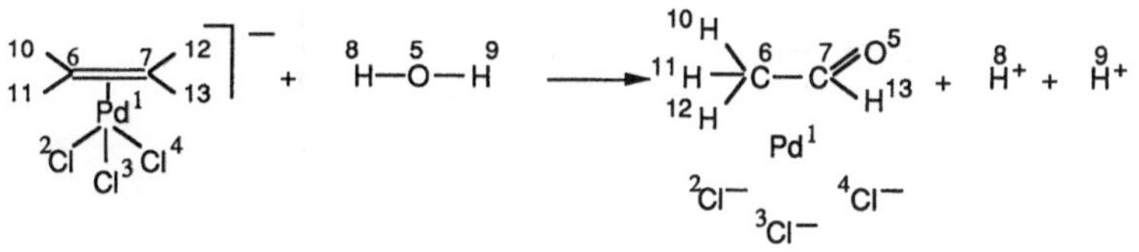

Fig. 2. The Wacker Reaction

This reaction is representable by the equation

$$XB + XR = XE$$

with

XB =

	1	2	3	4	5	6	7	8	9	10	11	12	13	DE1
1	6	1	1	1										1
2	1	6												
3	1		6											
4	1			6										
5					4			1	1					
6							1			1	1			1
7						1						1	1	1
8					1									
9					1									
10						1								
11						1								
12							1							
13							1							
DE1	1					1	1							4

Fig. 3. (see next page)

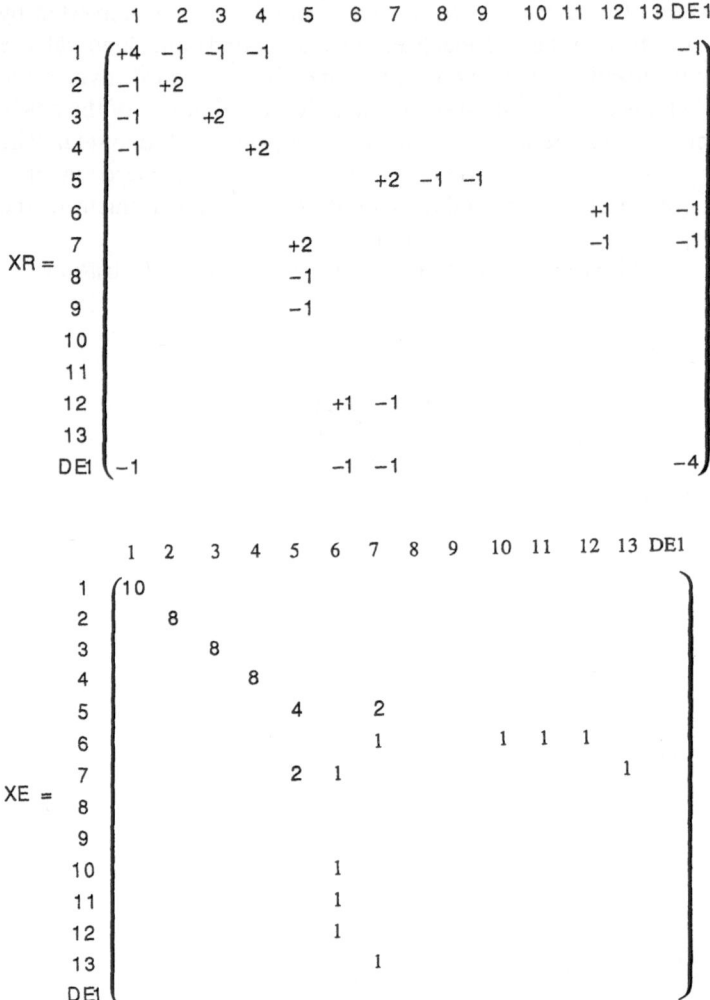

Fig. 3. Representation of the Wacker Reaction (only non-zero entries are shown)

The *xbe*- and *xr*-matrices belong to S(m + n), the additive group of all (m + n) × (m + n) symmetric matrices with integer entries, an (additive) free abelian group. Recall that the original *be*- and *r*-matrices belong to S(n). When the off-diagonal entries of the DE rows/columns are neglected, the *xr*-matrices belong to the group R(m + n) of (m + n) × (m + n) matrices with integer entries whose sum is zero.

Thus the algebras of the *be*- and *r*-matrices and the *xbe*- and *xr*-matrices [9] agree fully, as long as the chemistry-oriented boundary conditions are neglected.

As a general basis of computer programs for the solution of chemical problems, including their stereochemical aspects, a modification of the theory of the *xbe*- and *xr*-matrices and a corresponding data structure are very useful.

This data-structure [38] does not only convey certain information about the chemical constitution of the EM but also about its stereochemistry in terms of the theory of chemical identity groups [10].

The basic concepts of this data-structure and their respective characterization are now presented. In order to be able to describe molecular systems with electrons shared between more than two atoms, the following convention is introduced:

If electrons are delocalized among more than two atoms, i.e. more than one covalent bond, the corresponding bonds and atoms are treated as a DE-system. □

- EM

  An EM is uniquely determined by the following three types of information:
  - The set $A = \{A_1, ..., A_n\}$ of the constituent atoms.
    Each $A_i \in A$ is an atomic core. The FIEM(A), the set of all conceivable EM(A), is given by A.
  - The set of the chemical bonds in the EM.
    Each bond connects exactly two atoms A_i and A_j.
    The constitution of the given EM(A) is determined by the distribution of valence electrons.
  - The stereochemistry of the EM(A).
    The stereochemical information about a given EM(A) consists of one or more conceivable dissections of the EM into a skeleton and a set of ligands together with the respective chemical identity group, racemate group and constitution group [10].

- Atom

  Each atom $A_i \in A$ is described by the following data:
  - The unique atomic index of the atom in the set A.
    Such an index may be automatically assigned by a canonical ordering procedure such as the algorithm CANON [32].
  - The atomic number of the chemical element to which A_i is assigned.
  - The number of valence electrons that are not implicitly assigned to a particular chemical bond involving A_i.
    As a new feature, a separation of the lone valence electrons that are located directly at the atom A_i and the delocalized electrons that are shared by several atoms including A_i is introduced. Therefore three integer parameters are necessary:
    - The number of localized lone valence electrons.
    - The number of delocalized electrons in all DE-systems containing A_i.
    - The number of atoms in all DE-systems containing A_i.
  - The formal electrical charge.
    The treatment of the formal electrical charge is analogous to the treatment of lone and delocalized electrons:
    - The localized charge.
    - The delocalized charge in all DE-systems containing A_i.
    - The number of atoms in all DE-systems containing A_i.

- The number of implicit hydrogen atoms.
  This (optional) parameter may be used to account for those hydrogen atoms that are irrelevant for the considered EM and therefore shall not be represented by their own atomic descriptors.
- The set of "bondpartner-atoms" to which the atom A_i is connected through chemical bonds.

- Bond
  The chemical bonds of an EM are specified by three parameters:
- The two atoms that are connected through a given chemical bond.
- The type of the bond.
  The available bond types are defined in the following section.
- The set of bonds that form a DE-system. Depending on the type of the bond, this set may be empty. Each bond may only be contained in one DE-system.

4.1 Types of Chemical Bonds

One of the main differences between our proposed data-structure and the original concept of *be*-matrices is that there are no restrictions regarding the number of allowed types of bonds. Not only single, double or triple bonds are permitted. So far the following types have been included: *nonbonded, single, double, triple, quadruple, aromatic, three-center, double-three-center, bridge, hapto* and *connected*. If desired, other types can easily be defined and appended to this list, e.g. *intermetallic* or *cluster*.

The interpretations of the bond-types and the interplay between them as well as the concept of delocalized electron systems are described below.

For electron-counting purposes a well-defined number of valence electrons is assigned to the individual bond types:

type	implicit electrons
nonbonded	0
single	2
double	4
triple	6
quadruple	8
aromatic	2
three-center	0
double-three-center	0
bridge	0
hapto	0
connected	0

nonbonded indicates non-existence of a chemical bond. Naturally, no implicit electrons are assigned to bonds of this type. Bond-descriptors containing *nonbonded* as type may be removed from the data structure without any loss of information.

single, *double*, *triple* and *quadruple* represent ordinary covalent bonds with integer bond orders. The appropriate numbers of valence electrons are assigned to those types: two, four, six resp. eight electrons.

For aromatic compounds *aromatic* is introduced as an additional type of bond, whereas aromatic bonds were previously expressed in terms of alternating single and double bonds according to the VB-theory. The bond type *aromatic* is interpreted as follows: each bond of this type comprises exactly two electrons (the σ-electrons), whereas the π-electrons are treated as delocalized electrons, which appear in the appropriate *atom*-structures. *Aromatic* is therefore equivalent to *single* with regard to electron-counting purposes. All *bond*-structures, which belong to an aromatic system, are formally combined into an DE-system.

A similar interpretation applies to three-center-two-electron bonds: formally none of the 3 bonds contain electrons the two delocalized electrons appear in the respective *atom*-descriptors. Diborane may serve as an example:

Fig. 4. Diborane

The type of the B-B-bond is *double-three-center*, the type of the B-μH-bonds is *three-center*, and the type of the B-H$_t$-bonds is *single*. There are two DE-systems of three-center bonds, the B-B bond is described by two *bond*-structures with type *double-three-center*, each belonging to one of the *three-center*-DE-systems. The *atom*-descriptors for the μ-Hs contain the following entries concerning the electron distribution:

> localized electrons: 0
> delocalized electrons: 2
> number of atoms: 3 (B, H, B)

The entries in the two boron structures are:

> localized electrons: 0
> delocalized electrons: 4 (2 per three-center-system)
> number of atoms (in both DE-systems): 4

For example, a hydrogen bridge is considered as a DE-system of two bonds of type *bridge* (with no implicit electrons) and four delocalized electrons (the participating lone pair of the donor-atom is included in the delocalization).

The *hapto*-type is used for organometallic compounds, where delocalized hydrocarbon systems are coordinated to a metal-center. Formally, there are no electrons assigned to the *hapto*-bonds, but the metal-center is included in the number of atoms, among which the delocalized electrons (and the charge, if present) are distributed.

The bond-type *connected* may be used whenever a treatment of bonds with all participating electrons delocalized is adequate to the problem, e.g. as is the case with nonclassical carbocations [37]:

 **Fig. 5.** Norbonyl cation

The two dashed lines represent bonds of type *connected*, which form an DE-system. The electronic data for each of the three affected atoms is:

> localized electrons: 0
> delocalized electrons: 2
> number of atoms: 3

Any information that a constitutional formula may contain, can thus be expressed in terms of a corresponding modified *xbe*-matrix.

4.2 The Representation of Chemical Reactions by Modified *xbe*- and *xr*-Matrices

In the original *r*-matrices the entries are integer increments representing the changes in formal bond order, or the changes in the numbers of free electrons during the considered reaction. Thus, the additive transformation of the matrix B by R, according to B + R = E yields the *be*-matrix E of the product. Analogously chemical reactions are represented by XB + XR = XE in terms of the extended matrices.

A practice-oriented modification of the latter approach in order to also include the chemistry of molecules with non-integer formal covalent bond orders requires a mathematical device that is capable of taking into account chemical considerations beyond the formal, and of transforming the aforementioned data structures according to chemical reactions.

The addition of *xbe*- and *xr*-matrices is replaced by another composition R ⊕ B = E, that seems to be more complex, but is as elementary as the addition of integers. In the new formalism [38] the entries in the *r*-matrix (or *xr*-matrix) correspond to functions that act on the corresponding entries in the counterpart of a *be*-matrix (or *xbe*-matrix). As in the case of bond types, the number of allowable functions is not fixed. If desired, it is possible to append any number of additional functions. Thereby, new types of reaction mechanisms can be handled. However, there are chemical constraints regarding the applicability of a given function to the available types of chemical bonds and electron distributions, so we have partial functions. For example, in order to apply a function that acts on the three-center bonds of an EM, the latter must contain a system of three center bonds.

The various types of functions are the following:

The so-called *D-functions* concern the descriptors for the n atoms. As described above, in the representation of the electron distribution at any atomic center the localized and delocalized electrons are considered separately. In the *atom*-descriptor, the number of delocalized electrons is specified together with the number of associated atomic centers to which they belong. Therefore, any *D-function* acting on the atomic center with index i $(1 \leq i \leq n)$ should in principle consist of three components $(\Delta e_i.l \mid \Delta e_i.d \mid \Delta e_i.z)$. The first component $\Delta e_i.l \in \mathbb{Z}$ is interpreted as the increase or decrease of the number of localized electrons, the second component $\Delta e_i.d \in \mathbb{Z}$ as the increase/decrease of the total number of delocalized electrons, and the third component $\Delta e_i.z \in \mathbb{Z}$ represents the change in the total size of all DE-systems of which A_i is a member, in terms of the number of participating atoms.

In the modified *xr*-matrices $\Delta e_i.l$ corresponds to the diagonal entry $r_{i,i}$ $(1 \leq i \leq n)$. $\Delta e_i.d$ corresponds to the sum of the diagonal DE-entries $\sum_{1 \leq k \leq m} r_{n+k,n+k}$, with the further condition for k that the atom A_i is affected by the formation or break-up of the k-th DE-system, i.e. $r_{i,n+k} = r_{n+k,i} = \pm 1$.

$\Delta e_i.z$ is equal to the number of *different* atoms that are involved in the formation of DE-systems concerning A_i $(r_{i,n+k} = +1)$ minus the number of *different* atoms involved in the break-up of DE-systems concerning A_i $(r_{i,n+k} = -1)$.

The three components of the triplet are vectorially added to the corresponding entries of the *atom*-descriptor. However, this addition is subject to the limitation, that the resulting vector may not contain any negative elements.

The off-diagonal functions $r_{i,j}$ $(1 \leq i,j \leq n; i \neq j)$ act on bonds. Therefore they are called *B-functions*. So far the available set of *B-functions* BF is

$$BF = \{c,\ c^-,\ p1,\ p2,\ p3,\ p4,\ m1,\ m2,\ m3,\ m4,\ ar0,\ ar1,\ ar2,\ ar3,$$
$$ar0^-,\ ar1^-,\ ar2^-,\ ar3^-,\ tc0,\ tc1,\ dtc0,\ dtc1,\ tc0^-,\ tc1^-,$$
$$dtc0^-,\ dtc1^-,\ b0,\ b1,\ b0^-,\ b1^-,\ ha,\ ha^-\}\ .$$

Each function $f \in BF$ has the functionality

$$f:\ (Bond, \mathbb{N}) \rightarrow (Bond, \mathbb{N})$$

with $\mathbb{N}$ representing the number of free electrons, and is, as a rule, precisely defined for a characteristic set of bond types denoted by B_f.

Naturally, the invariancy of the total number of electrons during the course of a chemical reaction must be taken into account. For each function appearing in the *xr*-matrix the resulting change of the number of free electrons must be well-defined (e.g. electrons must be supplied in order to convert a bond of a low formal bond order into a bond of higher formal bond order. These electrons are either formally gained by applicating a suitable *B-function* (it must reduce a bond order) elsewhere in the EM or they are converted from free or delocalized electrons by applicating a corresponding *D-function*). It is therefore possible to check the *xr*-matrix with respect to the above invariancy by inspection. For the representation of chemical reactions, a suitable interplay of *D*- and *B-functions* is crucial.

For classical bonds of type *nonbonded (connected)*, *single*, *double*, *triple* or *quadruple*, eight functions are available: *p1* ("plus 1"), *p2, p3, p4*, which raise the given bond order by 1, 2, 3 resp. 4 and the functions *m1* ("minus 1"), *m2, m3* and *m4*, which decrease the bond order by 1, ..., 4.

$\forall x \in \mathbb{N}$:

$p1\ (nonbonded, x + 2) = (single, x)$ $B_{p1} = \{nonbonded, single, double, triple\}$
$p1\ (single, x + 2) = (double, x)$
$p1\ (double, x + 2) = (triple, x)$
$p1\ (triple, x + 2) = (quadruple, x)$

$p2\ (nonbonded, x + 4) = (double, x)$ $B_{p2} = \{nonbonded, single, double\}$
$p2\ (single, x + 4) = (triple, x)$
$p2\ (double, x + 4) = (quadruple, x)$

$p3\ (nonbonded, x + 6) = (triple, x)$ $B_{p3} = \{nonbonded, single\}$
$p3\ (single, x + 6) = (quadruple, x)$

$p4\ (nonbonded, x + 8) = (quadruple, x)$ $B_{p4} = \{nonbonded\}$

$m1\ (single, x) = (nonbonded, x + 2)$ $B_{m1} = \{single, double, triple, quadruple\}$
$m1\ (double, x) = (single, x + 2)$
$m1\ (triple, x) = (double, x + 2)$
$m1\ (quadruple, x) = (triple, x + 2)$

$m2\ (double, x) = (nonbonded, x + 4)$ $B_{m2} = \{double, triple, quadruple\}$
$m2\ (triple, x) = (single, x + 4)$
$m2\ (quadruple, x) = (double, x + 4)$

$m3\ (triple, x) = (nonbonded, x + 6)$ $B_{m3} = \{triple, quadruple\}$
$m3\ (quadruple, x) = (single, x + 6)$

$m4\ (quadruple, x) = (nonbonded, x + 8)$ $B_{m4} = \{quadruple\}$

Each appearing *nonbonded* may be substituted by *connected*. In fact, if a bond order is reduced to zero, the type changes into *connected*. Only if, after the application of the entire *xr*-matrix, this bond is not part of any DE-system, the bond is considered as non-existent (type *nonbonded*).

These functions may not be applied to other types of bonds, such as *aromatic* or *three-center*.

Another function is used to form bonds of type *connected* between atoms, which were not connected previously:

$c\ (nonbonded, x) = (connected, x)$ $B_c = \{nonbonded\}$
$c^-\ (connected, x) = (nonbonded, x)$ $B_{c-} = \{connected\}$

Aromatic bonds may undergo conversions to single, double and triple bonds and vice versa. Furthermore, cyclisation reactions of alkenes at a metallic center involve a formal transformation *nonbonded → aromatic*. Eight functions are available:

$ar0$ *(nonbonded,* x + 2*) = (aromatic,* x*)* $B_{ar0} = \{nonbonded\}$
$ar0^-$ *(aromatic,* x*) = (nonbonded,* x + 2*)* $B_{ar0-} = \{aromatic\}$
$ar1$ *(single,* x*) = (aromatic,* x*)* $B_{ar1} = \{single\}$
$ar1^-$ *(aromatic,* x*) = (single,* x*)* $B_{ar1-} = \{aromatic\}$
$ar2$ *(double,* x*) = (aromatic,* x + 2*)* $B_{ar2} = \{double\}$
$ar2^-$ *(aromatic,* x + 2*) = (double,* x*)* $B_{ar2-} = \{aromatic\}$
$ar3$ *(triple,* x*) = (aromatic,* x + 4*)* $B_{ar3} = \{triple\}$
$ar3^-$ *(aromatic,* x + 4*) = (triple,* x*)* $B_{ar3-} = \{aromatic\}$

The action of the functions *ar0, ar1, ar2* and *ar3* implies the formation of an (aromatic) DE-system, the appropriate change in the number of delocalized electrons must be denoted in the corresponding diagonal DE-elements of the *xr*-matrix. Analogously, if the functions *ar0⁻, ar1⁻, ar2⁻* and *ar3⁻* are used, the DE-system is dissected.

Three-center bonds can either be broken or converted into single bonds. The following functions are available:

$tc0$ *(nonbonded,* x*) = (three-center,* x*)* $B_{tc0} = \{nonbonded\}$
$tc0^-$ *(three-center,* x*) = (nonbonded,* x*)* $B_{tx0-} = \{three\text{-}center\}$
$tc1$ *(single,* x*) = (three-center,* x + 2*)* $B_{tc1} = \{single\}$
$tc1^-$ *(three-center,* x + 2*) = (single,* x*)* $B_{tc1-} = \{three\text{-}center\}$
$dtc0$ *(nonbonded,* x*) = (double-three-center,* x*)* $B_{dtc0} = \{nonbonded\}$
$dtc0^-$ *(double-three-center,* x*) = (nonbonded,* x*)*$B_{dtc0-} = \{double\text{-}three\text{-}center\}$
$dtc1$ *(single,* x*) = (double-three-center,* x + 2*)* $B_{dtc1} = \{single\}$
$dtc1^-$ *(double-three-center,* x + 2*) = (single,* x*)*$B_{dtc1-} = \{double\text{-}three\text{-}center\}$

In analogy to aromatic bonds, restrictions exist concerning the applicability of the functions. Here also a bond system is affected. Therefore, the number of bonds per DE-system (it will be three) must be taken into account. The proper adjustment of the delocalized electrons (DE-entries of the *xr*-matrix) must be ensured.

For the type *nonbonded* the above remarks regarding the interconvertibility of the types *connected* and *nonbonded* hold.

The symmetric cleavage of diborane (its structure has been described above) may serve as an example:

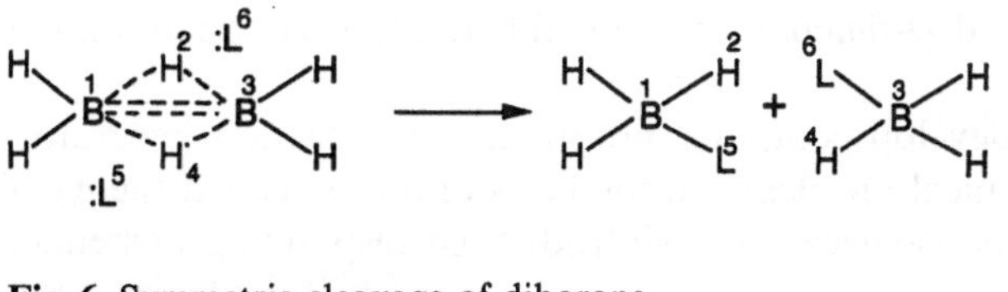

Fig. 6. Symmetric cleavage of diborane

XR-Matrix:

	1	2	3	4	5	6	DE1	DE2
1		$tc1^-$	$dtc0^-$	$tc0^-$	$p1$		-1	-1
2	$tc1^-$		$tc0^-$				-1	
3	$dtc0^-$	$tc0^-$		$tc1^-$		$p1$	-1	-1
4	$tc0^-$		$tc1^-$					-1
5	$p1$					-2		
6			$p1$				-2	
DE1	-1	-1	-1				-2	
DE2	-1		-1	-1				-2

Fig. 7. XR-matrix of the symmetric cleavage of diborane

For the treatment of *bridges* four functions are introduced:

$b0$ (*nonbonded*, x) = (*bridge*, x)　　　　　　$\mathbf{B}_{b0} = \{nonbonded\}$
$b0^-$ (*bridge*, x) = (*nonbonded*, x)　　　　　$\mathbf{B}_{b\,0-} = \{bridge\}$
$b1$ (*single*, x) = (*bridge*, x + 2)　　　　　　$\mathbf{B}_{b1} = \{single\}$
$b1^-$ (*bridge*, x + 2) = (*single*, x)　　　　　$\mathbf{B}_{b1-} = \{bridge\}$

As a prerequisite for the use of these functions a lone pair of electrons must be localized at one of the involved atoms, if a bridge is to be formed. The newly formed delocalized DE-system then comprises two bonds, four electrons and three atomic centers. The diagonal functions ensure the required electron balance.

The bond type *hapto*, that represents the coordinative bond of a delocalized hydrocarbon system to a metallic center, has two related functions:

ha (*nonbonded*, x) = (*hapto*, x)　　　　　　$\mathbf{B}_{ha} = \{nonbonded\}$
ha^- (*hapto*, x) = (*nonbonded*, x)　　　　　$\mathbf{B}_{ha-} = \{hapto\}$

If a *hapto* bond is formed, the metallic center is included in the delocalized coordination system, electrons and charge are now spread over the metallic center and the hydrocarbon system. Corresponding diagonal functions must be used.

Naturally, the functions described above can be combined. By applicating adequate sequences of *B*- and *D-functions* it is possible to represent an enormous variety of reactions.

In order to be able to develop computer programs with reaction generators [34] for the solution of chemical problems on the basis of this extended theory of the *xbe-* and *xr*-matrices, it is also necessary to introduce correspondingly extended transition tables [34].

4.3 Treatment of the Formal Electrical Charge

The application of any xr-matrix also influences the distribution of the formal electrical charge of each atom (localized) resp. DE-system (delocalized). The monitoring of thoses changes during the sucessive applications of the B- and D-functions is done automatically according to the following principles.

D-Functions

The triple $(\Delta e_i.l \mid \Delta e_i.d \mid \Delta e_i.z)$ (see 4.2) with

$\Delta e_i.l$: change in the number of localized electrons $\qquad$ $(\Delta e_i.l \in \mathbb{Z})$
$\Delta e_i.d$: change in the number of delocalized electrons $\qquad$ $(\Delta e_i.d \in \mathbb{Z})$
$\Delta e_i.z$: change of the number of atoms in the DE-system(s) $\qquad$ $(\Delta e_i.z \in \mathbb{Z})$

imply does not only the aforementioned changes of the corresponding data-triple for the description of the distribution of free and delocalized electrons at the atom i, but also of the data-triple concerning the formal electrical charge $(c_i.l \mid c_i.d \mid c_i.z)$ with

$c_i.l$: localized charge $\qquad$ $(c_i.l \in \mathbb{Z})$
$c_i.d$: delocalized charge $\qquad$ $(c_i.d \in \mathbb{Z})$
$c_i.z$: number of atoms in the DE-system(s) $\qquad$ $(c_i.z \in \mathbb{N})$.

In any case, $\Delta e_i.z$ is added to the value $c_i.z$:

$$c_i.z \text{ (new)} = c_i.z \text{ (old)} + \Delta e_i.z .$$

Further, $\Delta e_i.l$ is simply subtracted from the value $c_i.l$:

$$c_i.l \text{ (new)} = c_i.l \text{ (old)} - \Delta e_i.l .$$

If the xr-matrix contains DE-rows/columns, the formal charge of the concerned atoms is corrected for each such DE-row/column:

$$c_i.l \text{ (new)} = c_i.l \text{ (old)} - 1 \qquad (r_{n+k,n+k} > 0)$$

$$c_i.l \text{ (new)} = c_i.l \text{ (old)} + 1 \qquad (r_{n+k,n+k} < 0) .$$

If $r_{n+k,n+k} \neq \sum\limits_{1 \leq i \leq n} r_{i,n+k}$, a change of the delocalized charge results:

$$c_i.d \text{ (new)} = c_i.d \text{ (old)} + \left[\left(\sum\limits_{1 \leq i \leq n} r_{i,n+k} \right) - r_{n+k,n+k} \right] .$$

I. Ugi et al.

B-Functions

The influence of the various *B-functions* on the localized formal electrical charge of both atoms that are affected by a *B-function* may be summarized as follows:

$$c_i.1 \text{ (new)} = c_i.1 \text{ (old)} - (iE \text{ (new bond type)} - iE \text{ (old bond type)})/2$$

where iE symbolizes the number of electrons that are considered as inherent of a bond type (see page 15).

4.4 Access to the Data Structure

4.4.1 The Constitutional Part

The present concept requires a new, more efficient manipulation of EMs in the computer. An EM is described by three lists of essential information: a *list of atoms*, a *list of bonds* and a list of stereochemicaly relevant *partitionings*. All the elements of the lists are correlated through references with exactly defined semantics. Thus we have a dynamical structure without upper bounds for the number of atoms, or bonds. Furthermore, additional references between structural subunits join these directly in a chemically meaningful way, e.g. two *atom*-structures via the corresponding *bond*-structure and vice versa. This substantially facilitates the retrieval of information.

In Fig. 9 our approach is illustrated by a representation of ferrocene. In order to keep the example simple, only a few representative bonds concerning the atoms marked by numbers 1–4 are shown.

The iron ion Fe^{2+} carries 24 electrons. Six of these are lone valence electrons. In the list of electrons they are recorded as localized electrons. The 12 delocalized electrons of the two cyclopentadienyl-anions are included in the Fe-orbitals. Thus the electronic state of krypton ($36\,e^-$) is reached.

The central atom participates in two hapto DE-systems and appears electronically neutral. Six electrons of each cyclopentadienyl ligand are now distributed over six atoms, the delocalized negative charge disappears.

Each cyclopentadienyl-anion is an aromatic system and is treated separately. Each carbon atom participates in one aromatic and one hapto DE-system with six electrons delocalized over six atoms (five carbons and one iron).

Recall that there are always exactly two atoms involved in one bond. Accordingly, the iron atom has ten bonding partners. These bonds are of the "*hapto*" type.

Each carbon atom has one single bond to a hydrogen, two aromatic bonds to carbon atoms and one *hapto* bond to the iron atom. All carbon-carbon bonds of one ligand are concatenated into a contiguous system, just like the five hapto bonds between the central iron and the five carbon atoms of either ligand.

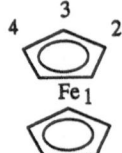

Fig. 8. Ferrocene

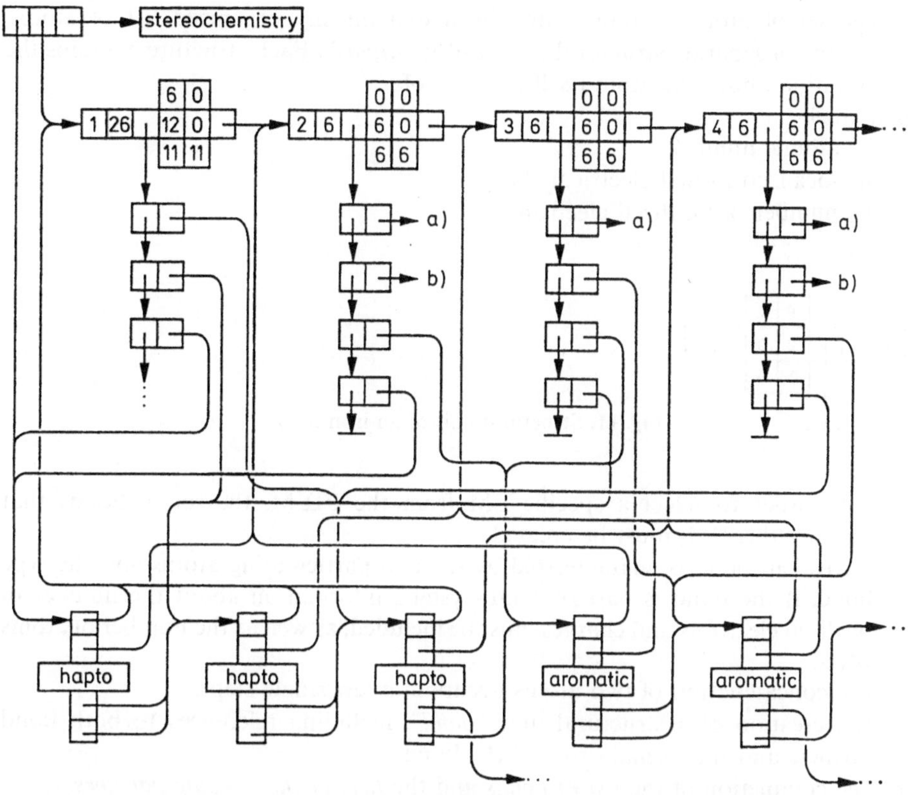

Fig. 9. Part of the data-structure of ferrocene. *a)* C−H-bond; *b)* C−C-bond

The use of this data structure for constitutional chemistry comprises three parts:
- representing an FIEM as a set of atoms
- representing an EM as a set of bonds between the atoms of an FIEM
- representing a chemical reaction and its action on an EM as a change of the set of bond orders of the affected EM.

1. Any EM is a representative of its FIEM; the simplest is the EM(A) of free atoms. Hence, the representation of an FIEM is accomplished by first defining a reference *ensemble* and then gradually generating the *list of atoms* belonging to that FIEM.

Ensemble references a structural unit including the starting elements of the list of atoms, the list of bonds and the stereochemistry (Fig. 10).

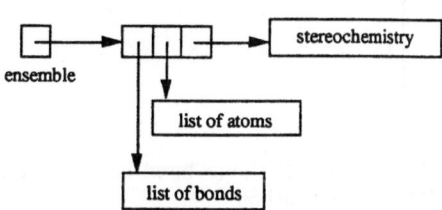

Fig. 10. "Head of an EM"

The set of atoms is represented by a dynamically organized list of atoms, consisting of separate structural units called *"atoms"*. Each structure contains the information about the atom itself in terms of

- a label,
- the atomic number,
- the localized formal electrical charge and
- the number of localized electrons.

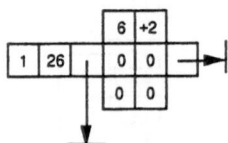

Fig. 11. Structural unit of an iron-ion

2. In order to select a specific EM from the FIEM, the set of bonds that determines this EM, must be created.

A covalent bond is characterized by the two participating atoms and the type of bond. If the bond is part of a DE-system, information about the number of delocalized electrons and charges must be included, as well as the number of atoms involved.

The concatenation of two atoms occurs in a few single steps.

- The creation of a structural unit *"bond"*, including references to both bond partners and the formal order of the bond.
- The completion of the *list of bonds* and the *lists of the bonding partners*.
- The incorporation of the bond into an existing delocalized DE-system, if necessary. This is accomplished by adjusting the reference to the bonds and correcting the entries of the list of electrons and the list of charges.

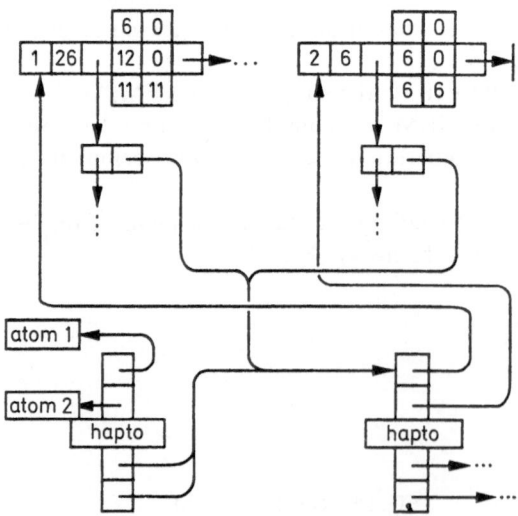

Fig. 12. Hapto bond between iron and carbon

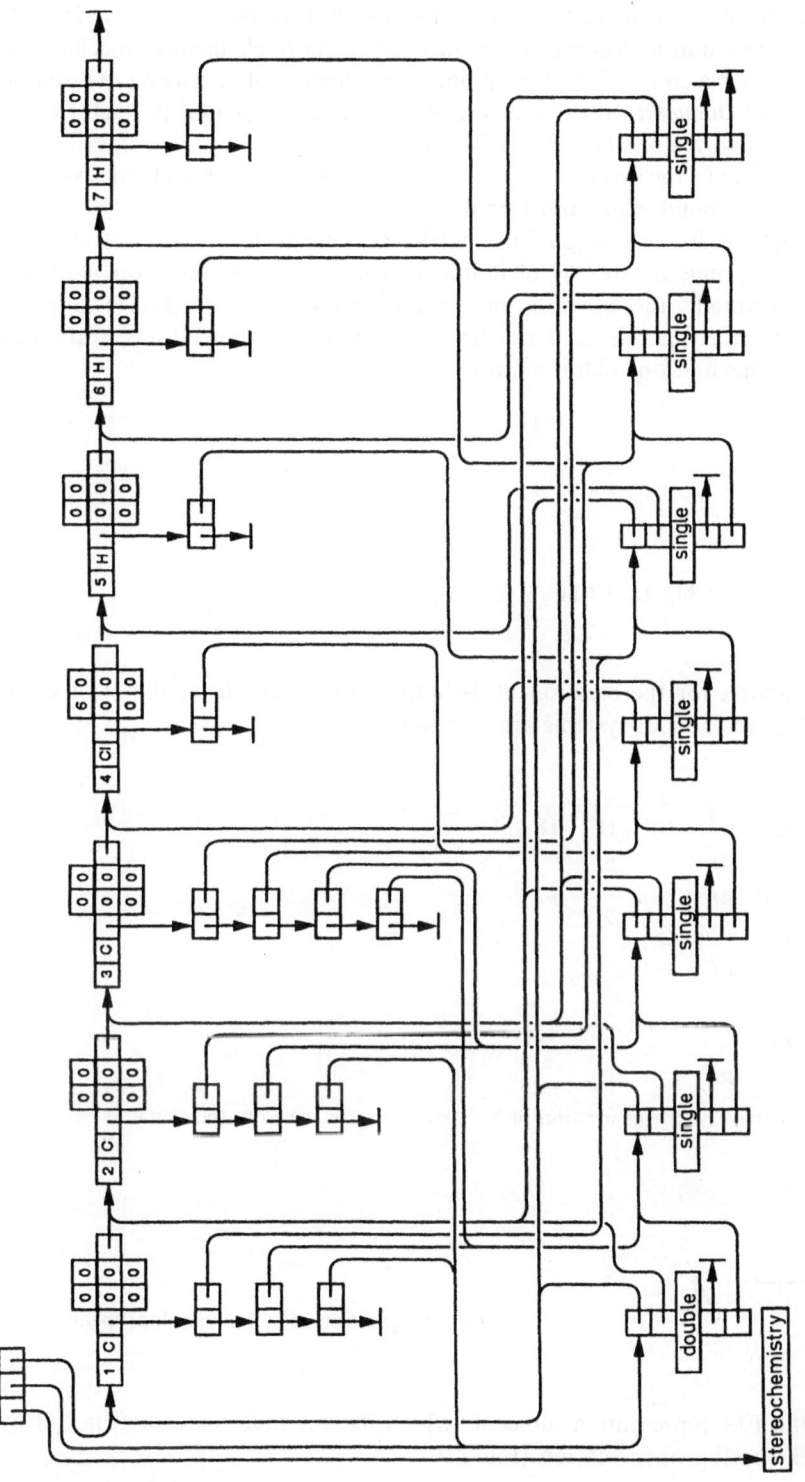

Fig. 14. Data-structure for 3-chlorocyclopropene

3. A chemical reaction is the conversion of an EM into an isomeric EM. Generally, a reaction is decomposable into one or more elementary mechanistic steps. Hence such a conversion corresponds to a sequence of intermediate isomeric EM's. *Ugi and Dugundji* call such a conversion a *reaction pathway*. Each step of the pathway involves *breaking* or *making a bond*.

Breaking a bond is the inverse of *making a bond* (as described above). The essential parameter is the bond which must be deleted.

Recall that the list of atoms whose entries all refer to localized electrons and charges corresponds to an EM of nonbonded atoms, i.e. a collection A of free atoms. Accordingly the creation of 3-chlorocyclopropene (C_3H_3Cl) (Fig. 13, Fig. 14, p. 27) corresponds to a reaction of three free carbon atoms, three free hydrogen atoms and one chlorine atom.

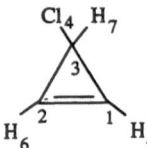

Fig. 13. 3-Chlorocyclopropene

The *xr*-matrix for the creation of 3-chlorocyclopropene from the elements is given in Fig. 15 (no DE-systems are necessary).

	C^1	C^2	C^3	Cl^4	H^5	H^6	H^7
C^1	-4	$p2$	$p1$		$p1$		
C^2	$p2$	-4	$p1$			$p1$	
C^3	$p1$	$p1$	-4	$p1$			$p1$
Cl^4			$p1$	-1			
H^5	$p1$				-1		
H^6		$p1$				-1	
H^7			$p1$				-1

Fig. 15. *xr*-matrix for the generation of 3-chlorocyclopropene from the elements

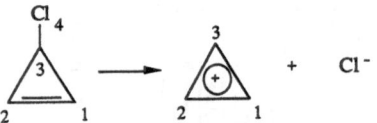

Fig. 16. Reaction of 3-chlorocyclopropene

Now, the EM represents a molecule which itself is able to react (Fig. 16) by forming the cyclopropene-cation (Fig. 17).

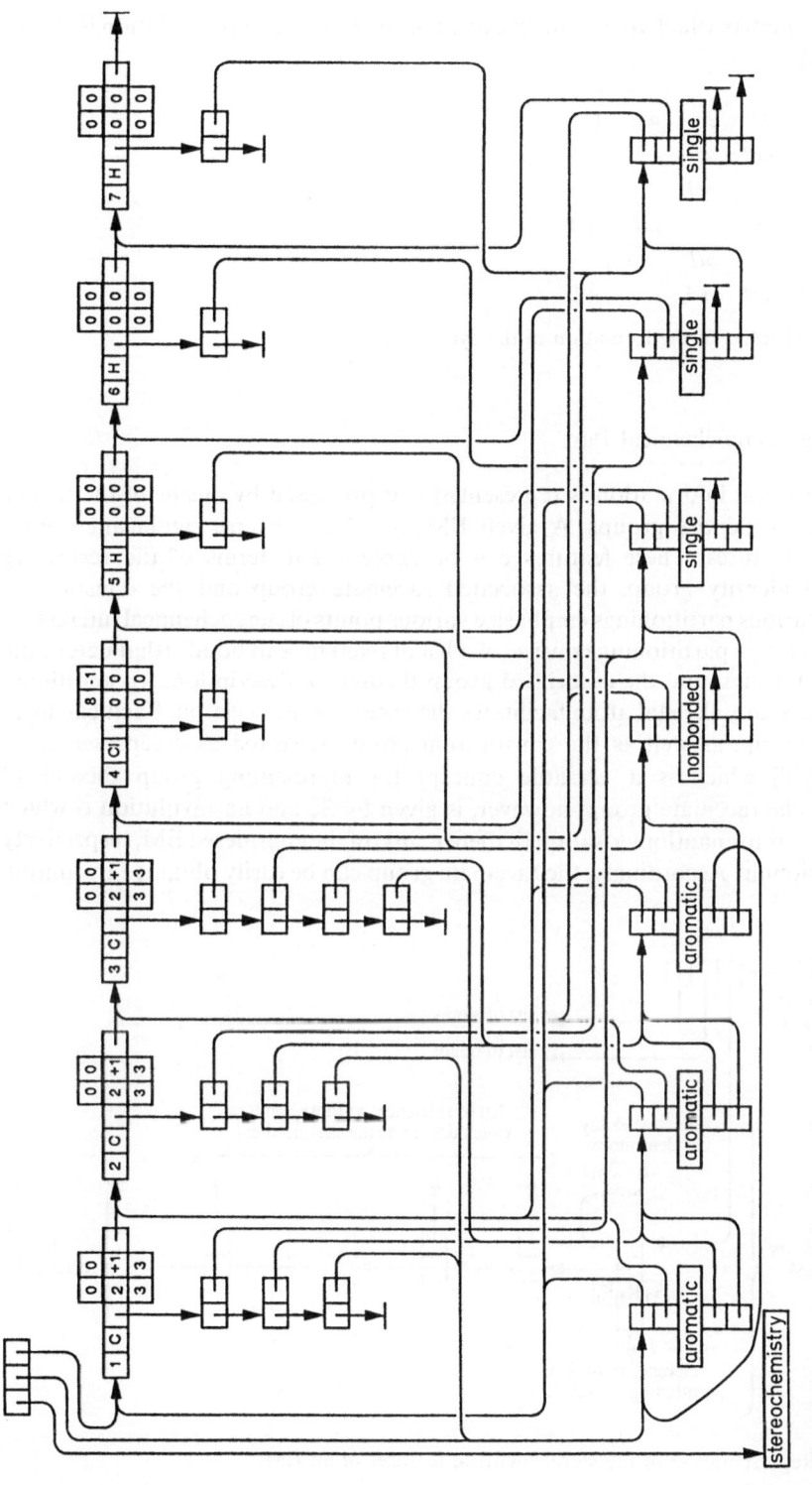

Fig. 17. Data-structure for the cyclopropyl-cation

The *xr*-matrix which represents the formation of the cyclopropyl-cation is shown in Fig. 18.

	1	2	3	4	DE1
1		ar2	ar1		+1
2	ar2		ar1		+1
3	ar1	ar1		m1	+1
4			m1		+2
DE1	+1	+1	+1		+2

Fig. 18. *xr*-Matrix of the formation of the cyclopropyl-cation

4.4.2 The Stereochemical Part

Stereochemical information is represented and processed by means of the theory of chemical identity groups. A given EM may have diverse significant stereo-chemical features. These features can be expressed in terms of the respective chemical identity group, the associated racemate group and the constitution group. Various partitionings emphasize various points of stereochemical interest.

In practice, a partitioning consists of a list of references to bonds (that determine the skeletal sites) and the associated group theoretical description. An additional reference to any skeletal atom facilitates the access to the skeleton. Each chemical identity group, as well as the constitution group, is stored as a representation matrix [40] which is a versatile concept for representing group theoretical objects. The racemate group, however, is given by S_x and an involution ϱ which corresponds to enantiomerisation (if there is any) of the considered EM, respectively its partitioning. Accordingly, the racemate group can be easily obtained by uniting

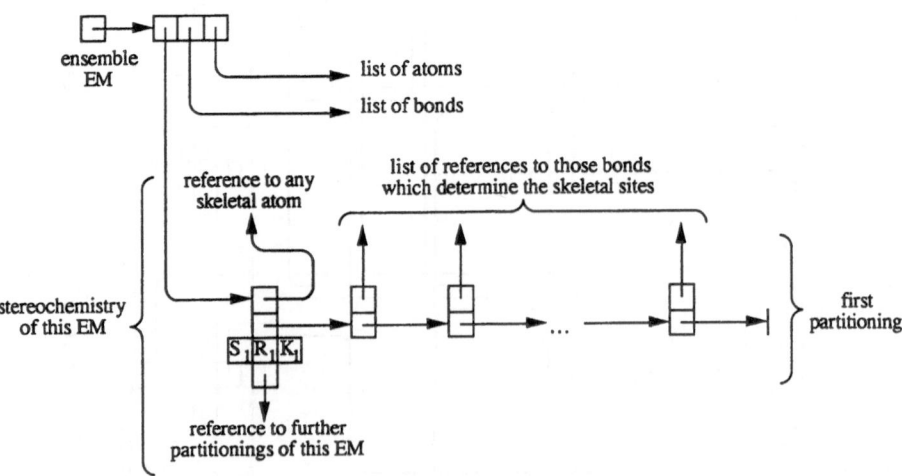

Fig. 19. Representation of the stereochemical features of an EM

union of the identity group S_X and its coset ϱS_X. The stabilizer group indicates the possible chemical identity of ligands and is obtained without user support.

In the example of 1-deutero-3-chlorocyclopropene two distinct partitionings serve for the stereochemical description of the EM: Partition 1 with the skeleton

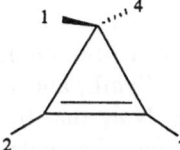

Fig. 20. 1-Deutero-3-chloropropene

and the set of ligands $L_1 = \{Cl, D, H, H\}$ has the chemical identity group $S_1 = \{\varepsilon, (14)\,(23)\}$, and is stored through the representation matrix

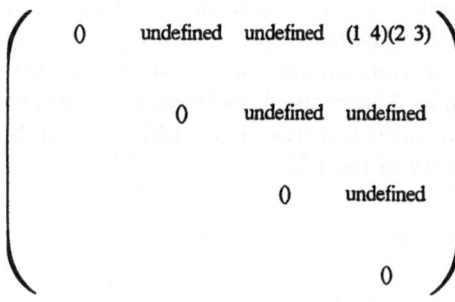

Fig. 21. Representation matrix for S_1

The racemate group $R_1 = \{\varepsilon, (1\,4), (2\,3), (1\,4)(2\,3)\}$, is stored in terms of the involution $(1\,4)$.

The constitution group K_1 is identical with the racemate group R_1; it is stored as a representation matrix:

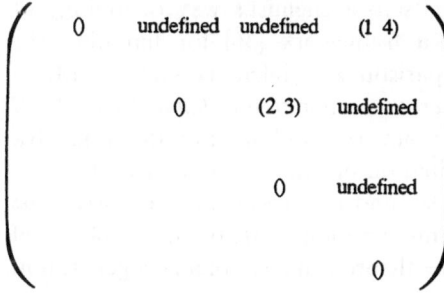

Fig. 22. Representation matrix for S_2

The stereochemical information that belongs to partitioning 2 with the skeleton

Fig. 23. Skeleton 2

and the set of ligands $L_2 = \{Cl, DC = ..., HC = ..., H\}$ differs from partitioning 1, belonging to the constitution group K_2. The latter is identical to SymL, and is therefore stored via the completely defined representation matrix. Note that the skeleton in partitioning 2 is not isomorphic to a tetrahedron, because the angles 1-C-4 and 2-C-3 differ.

The chemical equivalence of the ligands 3 and 4 in L_1 corresponds to the stabilizer group $\Sigma_1 = \{\varepsilon, (3\,4)\}$; the stabilizer group of the set of ligands L_2 is trivial.

This group theoretical information defines the equivalence classes of the various partitions. Each partitioning defines a family of permutation isomers. It is possible to correlate permutation isomers of distinct families as being chemically identical, which defines a further equivalence relation. Using this equivalence relation and the chemical identity group (or some other equivalence class) as an input for the versatile tool of set valued maps [41], a combination of the stereochemical information of partitioning 1 and partitioning 2 is obtained, and thereby — in case these partitionings represent the stereochemical features of the EM adequately and sufficiently — the entire stereochemistry of the EM is representable.

5 Conclusion

The present approach offers definite advantages for the computer manipulation of EMs:

- The inclusion of bonds with fractional bond orders extends the scope of the mathematical model of the constitutional chemistry from classical organic chemistry also to modern inorganic and organometallic chemistry.
- The direct combination of constitutional chemistry and stereochemistry affords the consideration of the stereochemical aspect in computer-assisted organic synthesis as well as in the analysis of structure/activity relations.
- The representation of an EM corresponds to a chemist's way of looking at molecules. Therefore we have created a framework [36] for handling the dynamical data structures without comparison with reference systems. Many procedures that exist in C and Pascal e.g. *create_atom*, *create_bond*, *delete_bond* facilitate handling. Thus, the creation of a new type of bond and the respective operator simply corresponds to a combination of some of these procedures.

The generalization of the theory of the *be-* and *r*-matrices now comprises the whole range of chemistry, and it takes into account both quantum chemical perspectives, VB and MO. It will serve as the theoretical basis of a new generation of chemical computer programs that will also be able to deal with the chemistry of electron deficient compounds of transition metals, including their stereochemical aspect.

Acknowledgement. We gratefully acknowledge the financial support of this work by the Volkswagen-Stiftung e. V. and the Fonds der chemischen Industrie e. V.

6 References

1. Dirac PAM (1929) Proc Roy Soc (London) 123: 714
2. Hjelt E (1916) Geschichte der Organischen Chemie, Vieweg & Sohn, Braunschweig
3. Cayley A (1857) Phil Mag 13: 19; (1874) 67: 144; (1857) Ber. 8, 1056
4. Balaban AT, ed (1976) Chemical applications of graph theory, Academic Press, London; see also Kvasnička V, Pospichal J (1990) Graph theoretical interpretation fo Ugi's concept of the reaction networks, J Math Chem 5: 309, and ref therein
5. Weyer J (1974) Hundert Jahre Stereochemie − Ein Rückblick auf die wichtigsten Entwicklungsphasen, Angew Chem 86: 604; (1974) Angew Chem Int Ed Engl 12: 591, and references therein
6. Polya G (1936) Algebraische Berechnung der Anzahl der Isomeren einiger organischer Verbindungen, Z Krystallogr (A) 93: 415; (1937) Acta Math 145; (1935) Compt Rend Acad Sci Paris 201: 1176; (1936) 202: 155; (1936) Vierteljschr Naturforsch Ges Zürich 81: 243
7. De Bruijn NG (1959) Koninkl Ned Akad Wetenshap Proc Ser A62: 59; in applied Combinatorial mathematics, Beckenbach EF ed (1964) Wiley, New York, p 144; (1970) Nieuw Arch Wiskunde (3) 18: 61
8. (a) Ruch E, Hässelbarth W, Richter B (1970) Doppelklassen als Klassenbegriff und Nomenklaturbegriffe für Isomere und ihre Abzählung, Theor Chim Acta 19: 288; Hässelbarth W, Ruch E (1973) Classification of rearrangement mechanisms by means of double cosets and counting formulas for the numbers of classes, Theor Chim Acta 29: 259; Hässelbarth W, Ruch E, Klein DJ, Seligman TH (1977) In: Sharp RT, Kolman B (eds) Group theoretical methods in physics, Academic, New York, p 617; (b) Ruch E (1972) Acc Chem Research 5: 49
9. Dugundji J, Ugi I (1973) An algebraic model of constitutional chemistry as a basis for chemical computer programs, Top Curr Chem 39: 19
10. Ugi I, Dugundji J, Kopp R, Marquarding D (1984) Perspectives in theoretical stereochemistry, Lecture Notes in Chemistry, vol 36, Springer, Heidelberg
11. Dugundji J, Kopp R, Marquarding D, Ugi I (1978) A quantitative measure of chemical chirality and its applications to asymmetric synthesis, Top Curr Chem 75: 165
12. Ugi I, Marquarding D, Klusacek H, Gokel G, Gillespie P (1970) Chemie und logische Strukturen (Chemistry and logical structures), Angew Chem, 82: 741; (1970) Angew Chem Int Ed Engl, 9: 703
13. Corey EJ, Wipke WT, Cramer RD, Howe WJ (1972) Computer-assisted synthetic analysis. Facile man-machine communication of chemical structure by interactive computer graphics, J Am Chem Soc 94/2: 421; Wipke WT, Braun H, Smith G, Choplin F, Sieber W (1977) SECS − Simulation and evaluation of chemical synthesis: Strategy and planning, in: Computer-assisted organic synthesis, Wipke WT, Howe WJ ed, ACS Symposium Series, p 98
14. Gelernter H, Sridharan NS, Hart HJ, Yen SC, Fowler FW, Shue HJ (1973) The discovery of organic synthetic routes by computer, Topics Curr Chem 41: 113
15. Ugi I (1969) A novel synthetic approach to peptides by computer planned stereoselective four component condensations of α-ferrocenyl alkylamines and related reactions, Rec Chem Progr 30: 389

16. Adler B (1986) Computerchemie — eine Einführung, VEB, Leipzig, chap 4
17. Ugi I, Bauer H, Brandt J, Friedrich J, Gasteiger J, Jochum C, Schubert W (1979) Neue Anwendungsgebiete für Computer in der Chemie, (new applications of computers in chemistry), Angew Chem 91: 99; (1979) Angew Chem Int Ed Engl 18: 111
18. Ugi I, Bauer J, Baumgartner R, Fontain E, Forstmeyer D, Lohberger S (1988) Computer assistance in the design of syntheses and a new generation of computer programs for the solution of chemical problems by molecular logic, Pure & Appl Chem 60: 1573
19. Ugi I, Wochner M, Fontain E, Bauer J, Gruber B, Karl R (1990) Chemical similarity, chemical distance and computer-assisted formalized reasoning by analogy, in: Concepts and applications of chemical similarity, Herausg.: Johnson MA, Maggiora GM, John Wiley & Sons, Inc, New York, p 239
20. Bauer J, Herges R, Fontain E, Ugi I (1985) IGOR and computer-assisted innovation in chemistry, Chimia 39: 43
21. Bauer J (1990) IGOR2: A PC-program for generating new reactions and molecular structures, Tetrahedron Comput Methodol 2: 269
22. Fontain E, Bauer J, Ugi I (1987) Computer-assisted bilateral generation of reaction networks from educts and products, Chem Letters, S. 37; Fontain E, Bauer J, Ugi I (1987) Computerunterstützte mechanische Analyse der Streith-Reaktion mit dem Programm RAIN, Z Naturforsch 42B: 889
23. Ugi I, Bauer J, Bley K, Dengler A, Fontain E, Knauer M, Lohberger S, (1991) Computer-assisted synthesis design, a status report, J Mol Structure (Theochem), 230: 73
24. Reitsam K, Fontain E, (1991) The generation of reaction networks with RAIN. 1. The reaction generator, J Chem Inf Comp Sci, 31: 96
25. Gasteiger J, Hutchings HG, Christoph B, Gann L, Hiller C, Löw P, Marsili M, Saller H, Yuki K (1987) A new treatment of chemical reactivity: Development of EROS, an expert system for reaction prediction and synthesis design, Top Curr Chem 137: 19
26. Ugi I, Bauer J, Fontain E (1990) Transparent formal methods for reducing the combinatorial abundance of conceivable solution to a chemical problem — computer-assisted eludication for complex reaction mechanisms, Anal Chim Acta 235: 155
27. Ugi I (1974) Eine neue Phase der Chemie. Logisches Strukturmodell und Computerprogramme zur Darstellung chemischer Verbindungen und Reaktionen, IBM Nachrichten 24: 180
28. Gasteiger J, Gillespie PD, Marquarding D, Ugi I (1974) From van't Hoff to unified perspectives in molecular structure and computer oriented representation, Topics Curr Chem 48: 1
29. Humboldt A v (1919) Chemiker Zeitung
30. Spialter L (1964) The atom connectivity matrix (ACM) and its characteristic polynomial (ACMCP), J Chem Doc 4: 261
31. Jochum C, Gasteiger J, Ugi I, Dugundji J (1982) The principle of minimum chemical distance and the principle of minimum structure change, Z Naturforsch 37B: 1205
32. Schubert W, Ugi I (1978) Constitutional symmetry and unique descriptors of molecules, J Am Chem Soc 100: 37; Schubert W, Ugi I (1979) Darstellung chemischer Strukturen für die computergestützte deduktive Lösung chemischer Probleme, Chimia 33: 183 see also Fontain E (1983) Qualitative Struktur-Wirkungs-Korrelation auf der Basis von Substrukturen an ausgewählten Beispielen, Diplomarbeit, Technische Universität München
33. Wochner M, Brandt J, Scholley A v, Ugi I (1988) Chemical similarity, chemical distance and its exact determination, Chimia 42: 217
34. Bauer J, Fontain E, Ugi I (1988) Computer-assisted bilateral solution of chemical problems and generation of reaction networks, Anal Chim Acta 210: 123; Ugi I, Bauer J, Fontain E (1990) Reaction pathways on a PC, in: Personal computers for chemists. J Zupan ed, Elsevier, Amsterdam, p 135
35. Ugi I (1986) Logic and order in stereochemistry, Chimia 40: 340; Ugi I, Gruber B, Stein N, Demharter A (1990) Set-valued maps as a mathematical basis of computer assistance in stereochemistry, J Chem Inf and Comp Sci 30: 485

36. Weidinger R (1991) Ein stereochemischer Netzwerkgenerator, Diplomarbeit, Universität Passau
37. Baeckvall JE, Akermark B, Ljunggren SO (1979) Stereochemistry and mechanism for the palladium (II)-catalyzed oxidation of ethene in water (the WACKER-process), J Am Chem Soc 101: 2411
38. Stein N (1990) Erweiterung der Theorie der BE- und R-Matrizen und Implementation der hierzu nötigen dynamischen Datenstruktur, Diplomarbeit, Technische Universität München
39. Olah G, Schleyer P von R (1972) Carbonium ions, vol 3, Wiley, New York; Bartlett PD (1965) Nonclassical ions; Benjamin WA, New York
40. Hoffmann CM (1982) Group-theoretic algorithms and graph isomorphisms, Lecture Notes in Computer Science, vol. 136, Springer, Heidelberg
41. Gruber B (1988) Die computerunterstützte Lösung stereochemischer Probleme auf der Grundlage der Theorie der chemischen Identitätsgruppen, Diplomarbeit, Universität Passau

Author Index Volumes 151–166

The volume numbers are printed in italics